Radio-Frequency Microelectronic Circuits
for Telecommunication Applications

Radio-Frequency Microelectronic Circuits for Telecommunication Applications

by

Yannis E. Papananos
*National Technical University of Athens,
Athens, Greece*

KLUWER ACADEMIC PUBLISHERS
BOSTON / DORDRECHT / LONDON

A C.I.P. Catalogue record of this book is available from the Library of Congress.

ISBN 978-1-4419-5104-5

Published by Kluwer Academic Publishers,
P.O. Box 17, 3300 AA Dordrecht, The Netherlands.

Sold and distributed in North, Central and South America
by Kluwer Academic Publishers,
101 Philip Drive, Norwell, MA 02061, U.S.A.

In all other countries, sold and distributed
by Kluwer Academic Publishers,
P.O. Box 322, 3300 AH Dordrecht, The Netherlands.

Printed on acid-free paper

Contents

Preface

The key target of telecommunication systems for the near future is "information available at any place, at any time and in every possible form". In order to achieve this target, a unified global system of personal and terminal stations must be available. Third generation mobile communication systems must provide global coverage and access to all basic and complementary services (e.g. PSTN, ISDN, B-ISDN). More specifically, these systems must aim to the integration of all available services into a unique system. The advent of satellite communication systems in 1997 is the first answer to this challenge. All major telecommunication players in Europe, Asia and the US have already put their systems in operation.

In all cases, low cost terminal devices are demanded in order to make the services accessible to the public. Driving demands of equal importance are the reduction of the terminal size and the increase in battery life. The average bill of materials for a mobile telephone is approximately 76 dollars today and it is estimated to drop to 58 dollars by the year 2000. An advanced GSM Phase 2 terminal weights 150 g and exhibits an average standby time of 350 h using lithium battery cells.

Contemporary mobile communication terminals use two or more integrated circuits in GaAs or Si technologies as well as a number of discrete components. The most critical part of these systems is the RF front-end since this is the part that mainly defines the quality of the transmitted/received signal as well as the levels of power consumption. In contrary to the baseband subsystem where the signal processing performed is mainly digital, the RF subsystem is purely analog. Today, the microelectronics industry concentrates on the design and development of integrated circuits operating in the radio-frequency range, using Si-based processes. Emphasis is put on CMOS technology due to its low cost compared to its bipolar counterparts. CMOS technology addresses the issue of RF analog and baseband digital circuit integration on the same silicon chip. This evolution will lead to a dramatic reduction in both cost and size of the terminal devices which are the main targets today: Only for the GSM system, the number of subscribers was 64 million in 1997 and is predicted to rise to 248 million by the year 2002.

The rapid developments in telecommunications in the past few years have led to a dramatic increase in the demand for RF integrated circuits, aiming at cost and power consumption reduction as well as at the step-up of the quality of services. The increased demand in RF ICs has inevitably led to a shortage in IC designers worldwide. The present book aims to cover certain aspects in silicon-based RF IC design. It is addressed to last-year undergraduate and postgraduate students in the field of microelectronic circuit design as well as to professional engineers that would like to approach the subject for the first time.

At this point, I would like to thank the Institute of Communications and Computer Systems of the National Technical University of Athens that supported the original edition of the present book in Greek, and personally, Prof. N. Uzunoglu, the Director of the Institute at that time. Mr. J. Finlay my publishing editor at KAP for entrusting me the English edition of this book. S. Bantas and Y. Koutsoyannopoulos for helping me in the preparation of the English edition. K. Frysiras for the LaTeX preparation of the original Greek edition of the book and last but not least, Prof. Y. Tsividis, Y. Koutsoyannopoulos, S. Bantas and A. Kyranas for their comments and suggestions.

July 1999

Yannis E. Papananos

CHAPTER
1

THE BIPOLAR TRANSISTOR AT HIGH FREQUENCIES

1.1 INTRODUCTION

All modern mobile communication systems continuously demand higher levels of integration in order to minimize the cost, reduce the terminal size and decrease the power consumption. For all systems operating up to 2 GHz (e.g. DECT wireless phones), integrated circuits for the RF, IF and baseband parts have already been fabricated and successfully used in commercial products. Moreover, one-chip solutions have appeared in the literature but up to now (1998), none of them has been applied in a commercial product. The basic cells of an RF transmitter are the Low Noise Amplifier in the receiver path, the Power Amplifier in the transmitter path and, the mixers and various filters in both parts. The most difficult part to integrate in a modern telecommunications receiver, is the RF front-end. Additionally, it is the part of the system that exhibits the highest power consumption. For this purpose, the selection of the proper technology for the integration of this particular subsystem plays a very important role. Among silicon processes, the bipolar technologies are the preferred ones.

The bipolar transistor exhibits two obvious advantages compared to the MOS transistor: a) higher transconductance for the same bias current (approx-

Technology	f_T (GHz)
Bipolar	25-50
BiCMOS	10-20
SiGe HBT	40-80

Table 1.1: Typical unity-gain frequency values of bipolar devices.

imately, two orders of magnitude higher) and b) higher unity-gain frequency. High values in transconductance, accomplishes the desired gain levels from the circuits without increasing the power consumption while the high unity-gain frequency permits the operation of the system at high frequencies. Three different technologies containing bipolar transistors exist today; in each of them, the performance of the bipolar transistor differs quantitatively. These technologies are: pure bipolar, BiCMOS and HBT SiGe. In BiCMOS technologies, the bipolar device co-exists along with the MOS device on the same Si substrate. There are obvious advantages here: since the RF subsystem is usually implemented using bipolar transistors and the baseband subsystem is a complex digital CMOS system, the BiCMOS technology allows for the complete integration of a mobile communications system on a single chip. Disadvantages of BiCMOS technologies are the increased process cost compared to a simple bipolar process and the reduced performance of the bipolar devices compared to their pure bipolar process counterparts. For example, the unity-gain frequency of the bipolar transistor in a pure bipolar process can be two times higher than of the bipolar transistor in a BiCMOS process. SiGe HBT technology allows for the implementation of both analog RF and digital systems at even higher frequencies of operation - close to the area of operation of optical communication systems. Again, the increased cost is the major disadvantage of the SiGe processes. In Table 1.1, typical unity-gain frequencies of bipolar transistors in the three above-mentioned processes are presented.

Recent advances in CMOS technologies and in particular the continuous shrinkage in transistors' channel dimensions (submicron technologies), makes the MOS device comparable to the bipolar transistor in terms of speed of operation. Moreover, considering that, under certain conditions, the MOS transistor is less noisy and more linear than its bipolar counterpart, it is likely that in the near future, when the MOS technology approaches the $0.1\,\mu$m barrier, it might become possible to fully integrate a commercial mobile communications system in a single chip. However, for the time being, the usage of bipolar transistors in the RF front-end part of telecommunications transceivers (especially in the superheterodyne architecture)is inevitable. For this purpose, in the present chapter, the structure of the bipolar transistor will be briefly reviewed in order for the reader to acquire an insight of the mechanisms that affect the operation of the device - especially at high frequencies. This way, the designer will be able to select the proper device for his application and fully exploit its merits for the benefit of the performance of the system under consideration.

The most popular type of bipolar transistors that it is used in RF IC design is the vertical *npn* device. In the current chapter, emphasis will be put in the presentation of the operation of this particular device but other types of transistors such as the lateral and vertical *pnp* transistors, will be also introduced.

1.2 THE STRUCTURE OF THE VERTICAL NPN BIPOLAR TRANSISTOR

The fabrication of an integrated bipolar transistor goes through a chain of consecutive process steps in order to develop all the necessary layers of materials that compose the device. The formation of the bipolar transistor starts from the common Si substrate which is usually a *p*-type semiconductor material. On top of this, a highly doped *n*-type region is diffused. This area will be used later on for the creation of the low-resistivity collector of the transistor and it is called *buried layer*. On top of this, an epitaxial layer is developed along the substrate. This epitaxial layer is a low-doped *n*-type material. The thickness and doping concentration of this layer along with the implementation of the base material, define the *breakdown voltage* (BV_{CBO}) of the transistor. The process if followed by two diffusion steps that create the *p*-type base and the *n*-type emitter of the device. A photochemical process is used to create holes in the insulating (SiO_2) material that is deposited on the Si surface in order to generate conductive paths via aluminum lines to the three terminals of the device, namely, the emitter, the base and the collector. In Fig. 1.1, a simplified view of the cross-section of the *npn* device is shown along with its schematic representation. In Fig. 1.2(a) the complete structure of an integrated bipolar transistor is shown and in Fig. 1.2(b) its layout is presented.

The ohmic contact between the collector terminal and the main body of the collector (epi layer, buried layer), is usually performed by diffusing a highly doped *n*-type material starting from the Si surface and deployed vertically down to the buried layer. This vertical region is called *sinker*. The combination of the sinker and the buried layer accomplishes the low collector resistance - an expedient property especially at high frequencies of operation.

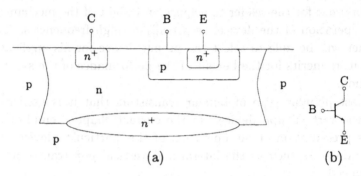

Figure 1.1: (a) Cross-section of an integrated bipolar npn transistor, (b) schematic representation of the bipolar npn transistor.

The major advantage of the IC design as compared to circuit design using discrete components is that the IC designer can modify the geometrical characteristics of the active and passive components in order to achieve optimized performance. Thus, the proper selection and design of the devices becomes an indivisible part of the circuit design process. A qualified microelectronic circuit designer must be able to properly evaluate the impact of device geometry to its electrical performance and properties. For instance, a transistor capable of providing high current levels is usually needed at the output stages of amplifiers. This can be easily achieved by increasing the geometrical dimensions of the device (bipolar transistor in our case). However, this leads to the inevitable increase in parasitic capacitances between base and emitter, collector and base and, collector and substrate; the frequency performance of the device deteriorates. This phenomenon must be taken into account during the design of the amplifier and particularly, with respect to the evaluation of its frequency response.

The basic electrical characteristics of the bipolar transistor, strongly related to its geometry are investigated next.

a) *Saturation Current I_S*. This is a basic parameter describing the performance of the bipolar transistor. It is given by the following expression [1]

$$I_S = \frac{qA\overline{D_n}n_i^2}{Q_B} \tag{1.1}$$

whereas A is the emitter-base junction area, Q_B is the total number of impurity atoms per unit area at the base of the transistor n_i is the intrinsic carrier concentration and $\overline{D_n}$ is the average effective value of the electron diffusion

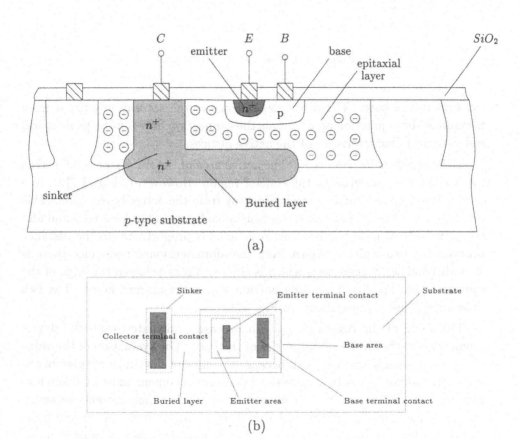

Figure 1.2: (a) Cross-section of an integrated bipolar *npn* transistor, (b) Layout of an integrated bipolar *npn* transistor.

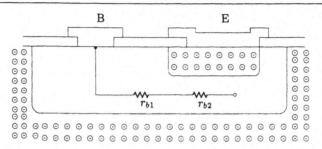

Figure 1.3: Base series resistance.

constant in the base. The direct dependence of the saturation current from the emitter-base junction area, reveals the connection between the geometrical and electrical characteristics of the active devices.

b) *Base Series Resistance* r_b. The active area of the transistor is defined as the base portion underneath the emitter region. However, in Fig. 1.2(a), it is shown that base terminal connector is away from the active base region. This topology introduces a notable series resistance between the base terminal and the actual active base region. This resistance is proportional to the distance between the two regions. Apart from the aforementioned resistance there is an additional series resistance which is the resistance between the edge of the emitter region and the base area portion where the current flows. The two resistors r_{b1} and r_{b2}, are shown in Fig. 1.3.

The value of the resistance r_{b1} can be easily calculated from the device geometry and the electrical properties of the base. The evaluation of the value of r_{b2} is not a simple task though because the current flow in this region of the transistor cannot be easily modeled and various phenomena must be taken into account - depending on the specific current value. For high currents however, the flow takes place mainly in the periphery of the emitter region and thus, the value of r_b approaches the one of r_{b1}. Due to this fact, whenever a low r_b value is demanded from the application (i.e. low-noise circuits), this can be achieved by increasing the perimeter of the emitter area close to the base terminal. Simultaneously, in order to keep the parasitic capacitances low, the emitter-base and collector-base junction areas must remain as low as possible. A good compromise is to form the emitter area by consecutive emitter fingers with base terminals laying in between.

c) *Collector Series Resistance* r_c. The collector series resistance r_c consti-tutes of three distinct equivalent resistors connected in series, as indicated in Fig. 1.4 : The resistance of the sinker area r_{c1} (resistance between the collec-tor terminal and the respective buried layer edge), the resistance of the buried

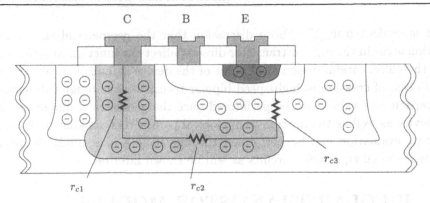

Figure 1.4: Collector series resistance.

layer r_{c2} and the resistance of the collector-base junction r_{c3} underneath the emitter terminal all the way to the buried layer edge. The exact calculation of the value of r_c (at least of parts r_{c1} and r_{c3}), is not an easy task to perform due to the complicated three-dimensional structure of the collector area. The simplest approximation for this area is that of the truncated four-side pyramid structure. It is also noted that the sinker area is not available in every bipolar technology but its existence helps reducing the value of r_{c1}.

d) *The Capacitances of the Bipolar Transistor*. The capacitances of the bipolar transistor are the two junction capacitances (emitter-base C_{je} and collector-base C_μ) as well as the capacitance between the collector and the substrate C_{cs}.

(i) *The emitter-base junction capacitance* (C_{je}). The emitter-base junction of a bipolar transistor is different than that of the simple pn junction due to the fact that the doping profiles in the two regions of the EB junction are different than that of the simple junction diode. Moreover, the carrier concentration in the lateral facets is a function of the distance from the surface and therefore, the analytical evaluation of this capacitance is very complicated. Thus, a numerical process is usually preferred for this calculation.

(ii) *The collector-base junction capacitance*. The collector-base junction capacitance comprises the area capacitance of the contact plane between base and collector and the respective sidewalls. The calculation of this capacitance follows that of the classical pn junction.

(iii) *The capacitance between the collector and the substrate*. This capacitance is consisted of three parts: the junction capacitance between the buried layer and the substrate, the lateral capacitance between the collector and the isolation walls in the substrate and, the capacitance between the epitaxial layer and the substrate.

It is evident from the above discussion that the geometrical size of the junction areas in the bipolar transistor directly affect the junction capacitances and, therefore, careful design and layout of the device is indispensable part in the process of creation of high-speed bipolar transistors. A rough estimate of the capacitance values can be performed once the geometrical dimensions of the device as well as the process technological parameters are known. A more accurate evaluation can be performed through SPICE-like simulators using the small-signal equivalent circuits as will be shown later on.

1.3   BIPOLAR TRANSISTOR MODELS

Extensive studies can be found in the relevant literature with respect to the operation and modeling of the bipolar device. In this section, a brief presentation of the small and large signal equivalent circuits will be performed. Emphasis will be put on the high-frequency operation as well as in the correlation of the above models to the models used by SPICE for integrated circuit operation simulations. Again, the presentation will be focused on the *npn* device. Similar analysis holds for the *pnp* device too.

The basic model used for the bipolar transistor is the Gummel-Poon model [2,3]. This is the model on which the existing SPICE model is based. It is an extension of another classical model, the Ebers-Moll model [4]. The latter is a non-linear dc model which has been modified in order to take into account various physical phenomena like the dependence of β on the collector current, the base width modulation etc.

1.3.1   The Small-Signal Model of the Bipolar Transistor

The existing large-signal models that describe the nature and the operation of the bipolar transistor under both static and dynamic conditions are quite complicated and non linear and therefore, cannot be used by the circuit designer for handy calculations. However, in many cases, the devices operate within a limited voltage and current range. Thus, when the transistor operates in small signal mode around its quiescent point, its non linear behavior can be approached by a simple linear model and hence, the computations become much simpler.

In analog electronic systems, the bipolar transistors are usually biased to operate in the active region. If the base-emitter junction voltage variations are small (quite smaller than the $\frac{kT}{q}$) value, then the transistor can be represented by an equivalent linear subcircuit. The $I - V$ equation that governs the operation of the bipolar transistor is repeated in Eq. (1.2).

$$I_C = I_S e^{qV_{BE}/kT} \tag{1.2}$$

The collector current variation is given by the derivative:

$$\frac{dI_C}{dV_{BE}} = \frac{qI_C}{kT} \equiv g_m \tag{1.3}$$

g_m is the *small signal transconductance* of the bipolar device. The variation of the base current is calculated in a similar manner:

$$\frac{dI_B}{dV_{BE}} = \frac{d(I_C/\beta_F)}{dV_{BE}} = \frac{g_m}{\beta_F} \equiv g_\pi \tag{1.4}$$

β_F is the current gain of the transistor under forward bias conditions. It is considered constant for the time being.

The collector current dependence on the base-collector junction voltage, describes the Early effect and is given by Eq. (1.5):

$$\frac{dI_C}{dV_{BC}} = \frac{I_C}{|V_A|} \equiv g_o \tag{1.5}$$

V_A is the Early voltage of the transistor. $r_o = \frac{1}{g_o}$ is the output resistance of the transistor.

The V_{BC} voltage variation changes the minority carrier concentration in the base and thus, directly affects the base current I_B. This phenomenon can be modeled by an ohmic resistor r_μ between the collector and the base of the transistor:

$$\frac{dI_B}{dV_{BC}} = \frac{g_o}{\beta_F} \equiv g_\mu \tag{1.6}$$

An additional aspect of the bipolar device behavior that is not predicted by the Ebers-Moll model, is the dependence of the base resistor from collector current variation. Thus, for the Ebers-Moll model we have $g_x = \frac{1}{r_b}$, where r_b is the series base resistance as defined in Section 1.2. In this model, a series collector resistance r_c (see Section 1.2) and a small series emitter resistance r_e are added. The latter can be ignored in the present model (its value is in the 1 Ω range) and it represents the emitter terminal contact resistance. Modern bipolar technologies implement high doping profiles in the emitter area in order to increase the value of β_F. However, it is worth noting that even a small-valued r_e results in a reduction of the base-emitter junction voltage by $r_e I_E$. This directly affects the base resistance (r_e is equivalent to a base resistance of $(1 + \beta_F)r_e$) and can lead to erroneous calculations.

Referring to Fig. 1.2(a) and the discussion in Section 1.2, the small-signal equivalent circuit model must also include the existing parasitic capacitances - especially if the device operates at high frequencies.

The collector-base junction capacitance is shown in the small-signal model as C_μ. For high reverse-bias junction voltages, the depletion region is mainly extended in the collector region, which maintains a uniform doping, profile. The capacitance C_μ is a non-linear function of the voltage across the terminals of diode BC.

The dependence of the above mentioned capacitance from the bias voltage, is given by Eq. (1.7):

$$C_\mu = \frac{C_{\mu 0}}{\left(1 - \frac{V}{\psi_0}\right)^n} \qquad (1.7)$$

where $C_{\mu 0}$ is the junction capacitance under zero bias voltage conditions (given as a technology parameter), n is a constant with value approximately equal to 0.5 and ψ_0 is the so-called built-in potential with a typical value of $0.55V$.

The capacitance between the collector and the substrate exhibits a similar behavior. However, it is noted that the substrate is usually bound to the most negative supply voltage of the circuit in order to ensure that all the isolation areas of the IC are separated by reverse-biased diodes. In that sense, for the small-signal operation of the circuit, the substrate is attached to the ground. Similarly, the bottom plate of the capacitor C_{cs} is always connected to ac ground. The value of C_{cs} for zero-bias voltage is given by a formula similar to Eq. (1.7).

Finally, another capacitance implemented in the $hybrid - \pi$ model of the bipolar transistor is C_π. This capacitance is connected in parallel to the conductance g_π mentioned earlier. C_π is constituted of two different parts: the emitter-base junction capacitance C_{je} and the base charging capacitance C_b. The two elements g_π and C_π are connected in between the emitter E and the internal base terminal B' of the transistor; that is the node following the base resistance r_b. The complete $hybrid - \pi$ small-signal equivalent circuit of the bipolar transistor is shown in Fig. 1.5.

The calculation of the exact value of capacitance C_{je}, is not an easy task to perform. A moderate value for this capacitance can be found by doubling

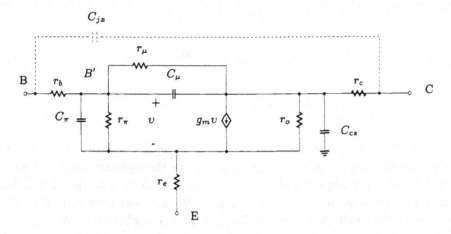

Figure 1.5: $Hybrid - \pi$ small-signal equivalent circuit.

the zero-bias capacitance C_{je0}. The latter is defined as a technology parameter, given by the silicon provider. The base charging capacitance is given by Eq. (1.8):

$$C_b = \tau_F g_m \tag{1.8}$$

τ_F is a quantity called base transit time in the forward direction. It defines the average time needed for a carrier to cross the base region in the forward direction. This time can be considered as a constant - independent of the operating conditions of the bipolar transistor and it is principally a function of the base width and the carrier concentration in the base region. The value of τ_F is a technology dependent parameter.

It is now evident from the previous discussion that the geometrical dimensions and the structure of the bipolar transistor drastically affect the electrical behavior of the device, especially at high frequencies of operation where the influence of parasitic capacitances prevails.

1.3.2 The Gummel-Poon Small-Signal Model

SPICE uses the Gummel-Poon model for the simulation of the operation of the bipolar transistor. The Gummel-Poon model is almost identical to the Ebers-Moll model with the addition of three elements that describe three more physical phenomena: The distributed capacitance between the base and the collector, the distributed element operation in the base region and the transit time τ_F modulation.

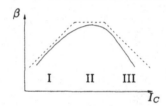

Figure 1.6: β dependence on I_C.

In the Gummel-Poon model, the distributed capacitance between base and collector (which plays a very important role in the operation of the bipolar transistor at high frequencies), is modeled by the addition of an extra capacitance C_{jx} between the collector and the *external* base terminal (Fig. 1.5). Details for the exact evaluation of C_{jx}, are given elsewhere ([1], [5]).

It was mentioned in Paragraph 1.3.1 that the base transit time τ_F can be considered constant in the Ebers-Moll model. In reality, this time increases for high collector current values and becomes a function of I_C and $V_{C}E$. This dependence can be described by a semi-empirical formulation. This phenomenon results in the reduction of the *unity gain frequency* f_T of the transistor which will be defined later on. The unity gain frequency of the transistor directly affects the maximum frequency of operation of the device and this is the main reason why in high-speed bipolar processes, the devices are optimized in terms of f_T for specific collector current densities.

Starting from fix-sized transistors, there are two ways to affect the geometrical dimensions and therefore the electrical behavior of the devices:

a) by connecting more than one identical devices in parallel and

b) by altering the emitter area (and thus the areas of all transistor regions) without affecting the structure of the device. Increase in area results in the increase of the saturation current I_s, the increase of parasitic capacitances and the respective decrease in the series base, collector and emitter resistances.

1.3.3   Dependence of Current Gain β on Collector Current I_C

It was mentioned earlier that the current gain of the transistor is not constant but it rather depends on the value of the collector current. Specifically, three distinct regions of variation are defined, as shown in Fig. 1.6: The weak current region where β increases with I_C, the intermediate range where β is almost constant and, the strong current range where β decreases with I_C.

The curve of Fig. 1.6 is extracted for constant V_{BC} voltage. The phenomena responsible for β variation are different in the three regions:

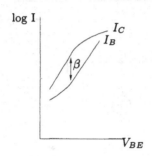

Figure 1.7: $\log I_C$ and $\log I_B$ characteristics as a function of V_{BE}.

Region I: Decrease in β values for weak currents (β_{FL}), is due to an additional component of the base current that is usually not taken into account. The variation of β is usually indirectly represented through the graph shown in Fig. 1.7. In this figure, the $\log I_C$ and $\log I_B$ variations on V_{BE} voltage are shown. Due to the logarithmic scale of the vertical axis, the value of β is directly given by the distance between the two curves.

Region II: In this region, the Ebers-Moll model is valid and β is called β_{FM}.

Region III: For high collector currents, the value of β (β_{FH})is drastically reduced as shown in Eq. (1.9) [1]:

$$\beta_{FH} \approx \frac{I_{SH}^2}{I_S} \beta_{FM} \frac{1}{I_C} \qquad (1.9)$$

where I_{SH} is the saturation current for high collector current values.

The reduction in β value is caused by carrier injection due to high current densities and also by the expansion of the base region in the collector region of the transistor. The latter is observed when the minority carrier concentration becomes comparable to the donor atom impurities (for a *npn* transistor) and it is called *Kirk effect*.

1.3.4   Frequency Response of the Bipolar Transistor

The parasitic capacitive elements of the small-signal equivalent circuit define the current gain of the transistor at high frequencies of operation. A quantitative measure of the high frequency performance of the bipolar device is the frequency at which the magnitude of the short-circuit current gain in a common emitter configuration becomes unity. This frequency is called *transition frequency* f_T and sets the highest possible frequency at which the transistor operates as an amplifier.

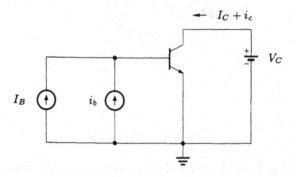

Figure 1.8: Circuit used in the evaluation of f_T

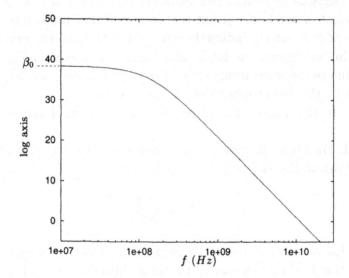

Figure 1.9: Small-signal current gain frequency response for a *npn* transistor.

The frequency f_T can be experimentally evaluated for each transistor or extracted from simulations once the technology parameters of the particular process are known. The circuit setup used in simulations for the calculation of the frequency response of the small-signal current gain of a bipolar transistor, is shown in Fig. 1.8.

The variation of β as a function of frequency for a bipolar transistor implemented in a BiCMOS process, is shown in Fig. 1.9.

Replacing the circuit of Fig. 1.8 by its Ebers-Moll small-signal equivalent circuit and after some calculations, Eq. (1.10) is derived. This equation gives a good approximation of the value of f_T:

$$f_T = \frac{1}{2\pi} \frac{g_m}{C_\pi + C_\mu} \qquad (1.10)$$

Considering the transistor in the BiCMOS process mentioned earlier and replacing in Eq. (1.10) the values of g_m, C_π and C_μ from the respective small-signal equivalent circuit, a value of $f_T = 11.2$ GHz is derived.

The single pole model of Eq. (1.10) is not adequate enough to describe the behavior of the device at high frequencies. In this case, non-quasi static effects in the operation of the transistor prevail. These phenomena can be taken into account if the transconductance gain of the voltage-controlled current source g_m^U of the Ebers-Moll model is replaced by a *transadmittance* $y_m \tilde{V}$ [6]. The current source is now dependent on the phasor $\tilde{V}$ of the voltage across r_π.

The transadmittance y_m is given by Eq. (1.11) [6]:

$$y_m = \frac{g_m}{1 + j(f/f_N)} \qquad (1.11)$$

where $f_N \approx 3(2\pi\tau_F)^{-1}$.

The single pole model accurately predicts the magnitude of the small-signal current gain of the bipolar transistor, even at high frequencies. However, the non-quasi static effects cause a phase lead, which is not predicted by Eq. (1.10), and this can generate an error in the evaluation of the phase of approximately $18°$ at the transition frequency. These phase errors can seriously affect the design of various types of circuits such as feedback amplifiers.

In Eq. (1.12), the accurate formula for the evaluation of f_T is given.

$$f_T = \frac{1}{2\pi\tau_{ec}} \qquad (1.12)$$

where

$$\tau_{ec} = \tau_E + \tau_B + \tau_C + \tau_C' \qquad (1.13)$$

τ_{ec} is the overall carrier transit time from the emitter to the collector. The terms in the right-hand part of Eq. (1.13), represent the intermediate transit times in between the various regions of the transistor and are mainly dependent on the geometry of the device.

Increase in f_T value can be achieved in various ways:

• Decrease in the width of the base region.

• Decrease in the collector area.

- Increase in the transistor's current (under certain conditions).

It is noted however that the decrease in the collector area results in a decrease of the breakdown voltage of the collector base junction which affects the operation of the transistor at higher supply voltages; this is usually not a problem in mobile communications applications. On the other hand, the increase in the collector current leads to an increase of the active base width. Kirk effect results in a peak f_T value for a given transistor geometry and given collector current. This observation leads to the formation of certain transistor families in every process - each exhibiting optimum high frequency performance for particular collector current magnitudes.

From the above discussion, it is evident that after all, the frequency f_T is not an adequate measure of the high frequency performance of the transistor. A more useful quantity can be the unity power gain frequency f_m or, otherwise, the maximum oscillation frequency:

$$f_m = \sqrt{\frac{f_{T0}}{8\pi r_b C_\mu}} \tag{1.14}$$

where f_{T0} is the maximum theoretically expected value of f_T. It is worth noting that transistors designed for maximum f_T do not necessarily exhibit optimum performance with respect to f_m. In Fig. 1.10, the dependence of f_T on the collector current for a particular npn device is shown. The transistor is designed for optimized performance when $I_C = 0.2$ mA.

1.3.5   The npn Bipolar Transistor in SPICE

SPICE-like simulators basically use models based on the Gummel-Poon model. A one to one correspondence of certain bipolar transistor parameters with their SPICE model counterparts, is shown in Table 1.2.

It is expedient to use a more accurate model of the bipolar transistor when this operates at high frequencies. Such a model is shown in Fig. 1.11.

As shown in the figure, the npn transistor is the internal transistor Q_{int}. In this transistor there are four terminals including the substrate terminal (SUB). The pnp transistor Q_{par} of the model is the parasitic transistor formed by the substrate region (p-type), the collector (n-type) and the base (p-type) of the transistor. A parasitic resistance r_{cv} is connected between the internal base terminal of the pnp transistor and the internal collector terminal of the npn transistor. This resistance represents a part of the ohmic losses in the collector region. In Fig. 1.12, a cross-section of an integrated vertical npn bipolar tran-

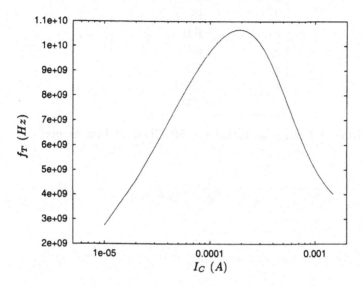

Figure 1.10: f_T as a function of I_C

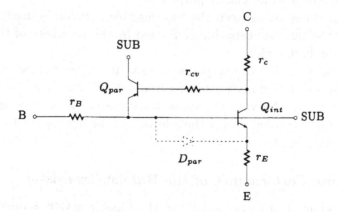

Figure 1.11: Equivalent circuit of the bipolar transistor at high frequencies

Parameter	SPICE Model
β	BF
I_S	IS
r_b	RB
r_c	RC
r_e	RE
C_{je}	CJE
τ_F	TF

Table 1.2: Bipolar transistor SPICE model parameters

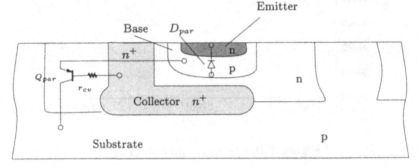

Figure 1.12: The parasitic elements of the vertical npn transistor

sistor over a p-type substrate is shown. All corresponding equivalent circuit elements are indicated for clarity purposes.

The D_{par} diode in between the base-emitter junction is used for better modeling of the junction capacitance formed by the periphery of the emitter area inside the base region.

It is noted that due to the parasitic resistances r_E, r_B and r_C, the actual npn transistor terminals exhibit lower voltages than the corresponding external terminals. This voltage drop on the parasitic ohmic resistors must be always taken into account during the selection of the biasing conditions of the transistors in a circuit.

1.3.6   Noise Performance of the Bipolar Transistor

Proper modeling of all noise sources in the bipolar device is imperative for high frequency applications since the minimization of noise is one of the most important steps in the course of an RF IC design.

The equivalent circuit model used for noise calculations in SPICE, is shown in Fig. 1.13.

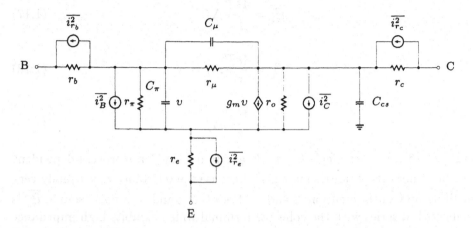

Figure 1.13: Equivalent noise model of the bipolar transistor

In Fig. 1.13, various noise sources are shown, as presented next:

Sources of thermal noise: These are the sources $\overline{i_{r_b}^2}$, $\overline{i_{r_c}^2}$ and $\overline{i_{r_e}^2}$ which refer to the three respective transistor terminal parasitic resistances. These parasitics are real ohmic resistors as opposed to resistances r_π, r_o and r_μ which are elements of the equivalent circuit *model* and thus, do not contribute to the thermal noise of the device.

The minority carriers crossing the base region towards the collector when the transistor is forward biased, generate shot noise in the collector region. This noise is represented by the noise source $\overline{i_c^2}$ in the circuit of Fig. 1.13. Similarly, the recombination that takes place in the base region as well as carrier injection from the base to the emitter, generate shot noise at the base. Moreover, it has been experimentally confirmed that at the internal node of the base-emitter junction, flicker and burst noise is generated. All the above three noise sources in the base-emitter area are represented in the circuit model by the noise source $\overline{i_B^2}$. All equations that define the noise sources of the bipolar transistors are given below:

$$\overline{i_B^2} = \underbrace{2qI_B\Delta f}_{\substack{\text{Shot}\\\text{Noise}}} + \underbrace{k_f\frac{I_B^{\alpha_f}}{f}\Delta f}_{\substack{\text{Flicker}\\\text{Noise}}} + \underbrace{\frac{k_bI_B}{1+(f/f_c)^2}\Delta f}_{\substack{\text{Burst}\\\text{Noise}}} \tag{1.15}$$

$$\overline{i_c^2} = 2qI_C\Delta f \tag{1.16}$$

$$\overline{i_{r_e}^2} = \frac{4kT}{r_e}\Delta f \tag{1.17}$$

$$\overline{i_{r_b}^2} = \frac{4kT}{r_b}\Delta f \tag{1.18}$$

$$\overline{i_{r_c}^2} = \frac{4kT}{r_c}\Delta f \tag{1.19}$$

In Eq. (1.15), the coefficients k_f, α_f, k_b and frequency f_c are process-dependent. The most important noise source is $\overline{i_{r_b}^2}$ because the resistance r_e is usually very small as previously mentioned and on the other hand, the noise source $\overline{i_{r_c}^2}$ is connected in series with the collector terminal which exhibits high impedance and therefore, can be ignored. A very important aspect in the evolution of a bipolar transistor aimed for RF applications is the minimization of its base resistance without degrading its high frequency performance.

Finally, it is worth noting that a very interesting and in-depth presentation of the noise performance of the bipolar transistor at high frequencies along with its implications in the design of low-noise amplifiers, can be found in [7].

1.4 OTHER BIPOLAR TRANSISTOR STRUC-TURES

In every integrated technology, the existence of a complementary element is imperative. Thus, a bipolar process must also contain a *pnp* device. It must be noted though that the process steps needed for the creation of such a device are completely different from that of its *npn* counterpart. The early bipolar transistor processes did not contain *pnp* devices. The performance of *npn* transistors in a modern bipolar process is mainly due to the existence of the epitaxial layer. Unfortunately, such a layer is not available for the *pnp* device, hence the latter exhibits poorer performance. However, the *pnp* device is a very crucial element in almost every integrated circuit: it is used in the design of biasing cells, as well as a load. These types of applications usually do not demand high performance from the device and therefore, the *pnp* transistor can be included in an RF IC design.

1.4.1 Lateral *pnp* Transistor

The simplest possible structure of a *pnp* transistor is the lateral structure shown in Fig. 1.14.

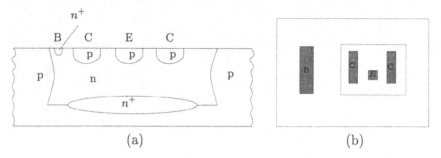

Figure 1.14: (a): Cross-section of a lateral *pnp* transistor. (b): Layout of a lateral *pnp* transistor

As shown in the figure, the collector region encloses the emitter while the base contact is outside. Current flow is as follows: holes are injected from the emitter region, flow across the silicon surface and are collected by the collector prior to reaching the base terminal contact. So in this case, the current flow is *lateral* as opposed to the vertical flow observed in the *npn* device.

The base region width is much bigger than in the *npn* transistor so that the depletion region can be kept away from the emitter when the collector-emitter voltage is maximized. Due to this fact, the transit time τ_F increases and therefore, the f_T frequency of the *pnp* device is much lower than that of the *npn* device (two orders of magnitude approximately).

Apart from the reduction in f_T, the current gain of the lateral *pnp* transistor is reduced as well: Minority carriers are injected from the base towards the emitter both edgeways and vertically. A fraction of these carriers is collected by the substrate which in this case plays the role of the collector of a parasitic vertical *pnp* transistor. Moreover, the emitter of the *pnp* transistor is more lightly doped than that of the *npn* device and in combination with the increased base width, the carrier injection capability is reduced. All the above result in a reduction of the β of the lateral *pnp* transistor which is further reduced as the collector current increases. A typical value for the β of the lateral *pnp* transistor is in between 30 and 50 for collector currents of up to several tenths of μ A. Above these current levels, the value of β is continuously reduced and becomes 10 or smaller for a collector current of approximately 1 mA.

1.4.2 Substrate *pnp* Transistor

An alternative *pnp* transistor structure is derived if the substrate is used as the collector of the device. The formation is shown in Fig. 1.15.

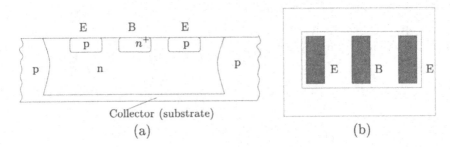

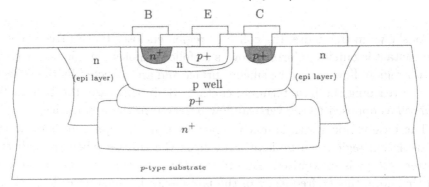

Figure 1.15: (a): Substrate *pnp* transistor structure (cross-section). (b): Substrate *pnp* transistor structure (layout).

Figure 1.16: Structure of a vertical *pnp* transistor

The major advantage of this structure is that the current flow is now vertical and the active emitter area is larger than that of the lateral *pnp* device. Thus, the substrate *pnp* transistor exhibits an improved current driving capability. Other than that, its performance is similar to that of the lateral *pnp* device because in both cases, a very important parameter, the effective base width, is large. The substrate *pnp* transistors are usually utilized in emitter-follower configurations.

1.4.3   Vertical *pnp* Transistor

In certain technologies (mainly BiCMOS), it is possible to implement vertical *pnp* bipolar devices, similar to the *npn* ones. The structure of such a transistor, is shown in Fig. 1.16.

In the vertical *pnp* transistor of Fig. 1.16, the collector is formed by a *p*-type well and a *p*+ doped buried layer. This structure favors the reduction of the collector series resistance. The collector is vertically isolated from the

p-type substrate through a $n+$ doped buried layer and horizontally, through the n-type epitaxial layer.

Starting from a BiCMOS technology, it is relatively easy and cost-effective to produce vertical pnp devices. The major advantage in this case is that the f_T frequency can easily go beyond 500 MHz and thus, these transistors can be effectively employed in various telecommunications and other high-frequency applications.

Bibliography

[1] P.R. Gray and R.S. Meyer, *Analysis and Design of Analog Integrated Circuits*, John Wiley, 1992.

[2] H.K.Gummel and H.C. Poon, "Modeling of Emitter Capacitance," in *IEEE Proc. (Lett.) 57*, 1969.

[3] H.K.Gummel and H.C. Poon, "An Integral Charge Control Model of Bipolar Transistors," *Bell Syst. Tech. J.*, vol. 49, 1970.

[4] J.J Ebers and J.L. Moll, "Large-Signal Behaviour of Junction Transistors," in *Proc. IRE*, 1954, vol. 42.

[5] D.J Roulston, *Bipolar Semiconductor Devices*, Mc Graw-Hill, 1990.

[6] Y.P. Tsividis, *Mixed Analog-Digital VLSI Devices and Technology*, Mc Graw-Hill, 1996.

[7] S.P. Voinigescu, M.C. Maliepaard, J.L. Showell, et al., "A Scalable High Frequency Noise Model for Bipolar Transistors with Application to Optimal Transistor Sizing for Low-Noise Amplifier Design," *IEEE J. of Solid-State Circuits*, vol. 32, no. 9, pp. 1430–1439, September 1997.

CHAPTER
2

THE MOS TRANSISTOR AT
HIGH FREQUENCIES

2.1 INTRODUCTION

Modern designs of telecommunications transceivers employ different technologies (GaAs, BiCMOS/bipolar, CMOS) depending on the frequency of operation of each stage. Evidently, this practice dramatically increases the system cost but it is a one way solution in many cases. However, at the low end of the frequency band (800 to 2400 MHz), it is now feasible to implement the complete system using Si-based technologies. This fact, drastically reduces the manufacturing cost and increases the integration level - a very desirable feature especially in mobile communication systems. Further cost reductions are possible if it becomes feasible to design and implement the complete transceiver in a pure CMOS technology which is the less costly solution and simultaneously, exhibits the highest level of integration. This solution is investigated in research projects from the Academia but the Industry still remains reluctant in adopting it mainly due to the fact that the MOS transistor exhibits poorer performance than its bipolar counterpart, especially at high frequencies of operation.

There are two ways to overcome the intrinsic problems of the MOS device: The adoption of alternatives to the classical superheterodyne architecture in

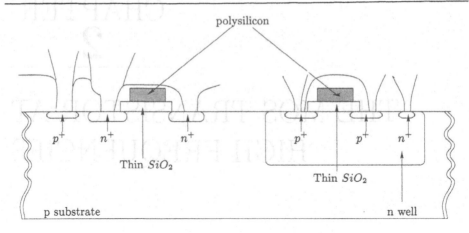

Figure 2.1: Cross-section of a typical CMOS technology

the design of the transceivers (e.g. direct conversion) and the usage of the most modern submicron CMOS technologies. Today (1998), $0.25\mu m$ CMOS technologies are available in the market while $0.1\mu m$ technologies are on their way [1-2]. In a $0.1\mu m$ technology, the unity gain frequency f_T of the MOS device can reach the 100 GHz range! Of course, there is always the problem of the low transconductance g_m value of the MOS transistor as opposed to that of its bipolar counterpart for a given current value as well as a number of other issues that will become evident later on.

2.2   THE MOS TRANSISTOR STRUCTURE

Early MOS transistor technologies could only implement pMOS devices. Next came the nMOS technologies in which only the n-channel device could be fabricated. Almost simultaneously with the nMOS technology, the CMOS technology appeared. The latter allows for the fabrication of both types of MOS devices on the same Si substrate. n-channel devices are approximately three times faster than their p-channel counterparts and this is due to the higher mobility of the electrons as opposed to that of the holes. In a BiCMOS technology is also possible to have both types of MOS devices. In Fig. 2.1, a cross-section of a typical CMOS technology is shown.

As shown in Fig. 2.1, the p-channel device is developed within a n-well. The n-well CMOS technology is the most popular CMOS technology today. The source and drain areas of the pMOS transistor are formed using a highly doped p material. The usage of the n-well, allows for the isolation of the bulk of each pMOS transistor and the respective formation of a separate terminal.

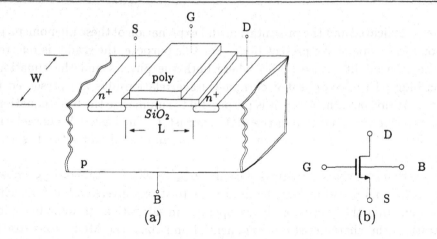

Figure 2.2: The nMOS transistor a) Physical layout b) Circuit representation

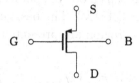

Figure 2.3: Circuit representation of the pMOS transistor

This way, it is possible for the body of each device to be biased separately, a property that is not encountered in nMOS devices. The latter share a common body which is the silicon p type substrate - always connected to the most negative supply available. In Fig. 2.2(a), the detailed formation of a nMOS device is shown. The gate (G) of the transistor is fabricated using a polysilicon layer. The gate is separated from the substrate (body - B of the transistor) by a thin SiO_2 layer. The source (S) and drain (D) areas are formed by a highly doped n-type material within the p-type substrate.

The transistor channel is formed by the substrate region underneath the gate and between the source and drain areas. As can be seen in Fig. 2.2(a), the channel length is L and the channel width is W. The geometrical dimensions of the channel are a very important parameter that define the electrical behavior of the device as will be shown later on. This property is a very powerful tool in the hands of the IC designer. The oxide thickness between the gate and the channel has a typical value between 20 and 200 $\overset{o}{A}$ in a modern VLSI technology. The pMOS devices are constructed in a similar manner. In Fig. 2.3, the circuit representation of the pMOS transistor is shown.

The mechanisms that determine the operation of the MOS transistor are quite complicated and the presentation and explanation of these phenomena go beyond the scope of the present book. For this purpose, the reader is referred to the relevant literature ([3-5]) where both the physics and the equations describing all the complicated current-voltage relationships are presented in detail. At this point however, it is purposeful to summarize the basic equations that describe the operation of the MOS transistor. The basic I-V characteristics of the MOS device are shown in Fig. 2.4(a) and the circuit setup is shown in Fig. 2.4(b).

A very useful characteristic is the curve of $\sqrt{I_D}$ as a function of V_{GS}, shown in Fig. 2.5 for V_{GS} values ranging from weak to strong inversion but for such a V_{DS} value that the transistor always operates in saturation. In weak inversion operation, the characteristic is exponential and thus, the MOS transistor in this case exhibits a behavior similar to the bipolar transistor. Due to the low value of V_{GS} in weak inversion, this mode of operation is preferred in extremely low consumption applications (LP-LV). The basic equations that describe the operation of the device in this region are presented next:

$$I_D = I_M \exp\left(\frac{V_{GS} - V_M}{n\phi_t}\right)\left[1 - \exp\left(-\frac{V_{DS}}{\phi_t}\right)\right] \qquad (2.1)$$

where

$$I_M = I'_M \frac{W}{L} \qquad (2.2)$$

and

$$I'_M = K'(n - 1)\phi_t^2 \qquad (2.3)$$

$$n = 1 + \frac{\gamma}{2\sqrt{V_{SB} + \phi_0}} \qquad (2.4)$$

$$\gamma = \frac{\sqrt{2q\epsilon_s N_B}}{C'_{ox}} \qquad (2.5)$$

$$K' = \mu C'_{ox} \qquad (2.6)$$

$$\phi_t = \frac{\hat{k}T}{q} \qquad (2.7)$$

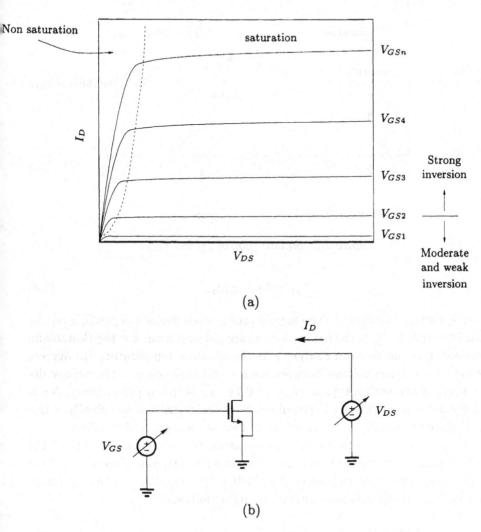

Figure 2.4: a) I-V characteristics of the nMOS transistor b) Circuit setup

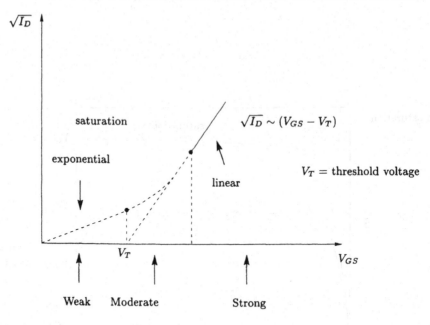

Figure 2.5: Square root of I_D vs. V_{GS}

$$V_M = V_T - cn\phi_t \qquad (2.8)$$

μ is the carrier mobility in the channel (holes or electrons - depending on the transistor type), C'_{ox} is the oxide capacitance per unit area, $\hat{k}$ is the Boltzmann constant, q is the electron charge, T is the absolute temperature (in degrees Kelvin), V_{SB} is the voltage between source and bulk, ϕ_0 is a technology dependent voltage with a typical value of $0.7V$. ϵ_S is the Si permittivity, N_B is the doping concentration in the substrate of the transistor and finally, c is a small number (usually between 1 and 3) that is technology dependent.

There is no simple formulation expressing the I-V characteristic of the MOS transistor operating in the moderate inversion region. This is one of the most serious problems encountered in SPICE-like simulators. The equations that hold in strong inversion operation, are as follows:

$$I_D = \frac{1}{2}\frac{\mu C'_{ox}}{1+\delta}\frac{W}{L}(V_{GS} - V_T)^2 \qquad V_{DS} \geq V'_{DS} \quad \text{(saturation)} \qquad (2.9)$$

$$I_D = \frac{1}{2}\mu C_{ox}{}'\frac{W}{L}[2(V_{GS} - V_T)V_{DS} - (1+\delta)V_{DS}{}^2]$$
$$\text{if } V_{DS} \leq V'_{DS} \quad \text{(non saturation)} \qquad (2.10)$$

where

$$V_T = V_{TO} + \gamma(\sqrt{V_{SB} + \phi_o} - \sqrt{\phi_o}) \qquad (2.11)$$

where V_{TO} is the threshold voltage for zero V_{SB} value (given in a certain technology). δ is a technology dependent constant; its value can be consider zero in first-order approximations.

$$V'_{DS} = \frac{V_{GS} - V_T}{1 + \delta} \qquad (2.12)$$

The dependence of I_D from V_{DS} in the saturation region is taken into account as follows:

$$I_D = I_D'\left(1 + \frac{V_{DS}}{V_A}\right) \qquad (2.13)$$

where I_D' is given by Eq. (2.9). This is equivalent to the Early effect of the bipolar transistors and the voltage V_A is proportional to the length of the channel of the MOS device and it is a technology dependent parameter.

Finally, it is worth noting that the MOS transistor enters the strong inversion operation when the V_{GS} voltage is higher than the threshold voltage V_T by approximately 200 mV. The threshold voltage depends on the particular technology and the trend is to reduce it as much as possible for low-voltage circuit operation. Contemporary MOS technologies have entered the submicron lithography and, particularly in deep submicron technologies, the V_T value can be as low as 0.1 V for voltage supplies as low as 1 V [6].

The previously presented equations, describe the dc operation of the MOS transistor. However, in the present book, we are mostly interested in the frequency-dependent operation of the device biased in the strong inversion (saturation or non-saturation) region.

2.3 MOS TRANSISTOR SMALL-SIGNAL MODELS

2.3.1 Small-Signal Equivalent Circuit at Low Frequencies

Having defined the bias conditions of the MOS device as shown in Fig. 2.6, we start varying the voltages V_{GS}, V_{SB} and V_{DS} by a small fraction - one at a time. For each variation, its influence in the drain current I_D is inspected. For example, V_{GS} is varied by ΔV_{GS} and the new drain current value is examined.

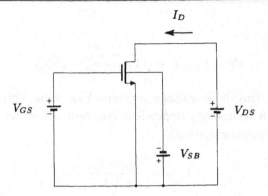

Figure 2.6: nMOS transistor biasing

The new value for the drain current is calculated in dc steady-state. The quantity of interest is the ΔI_D variation between the initial and final steady-state conditions. These current variations can be expressed by the following conductances:

a) Gate transconductance g_m:

$$g_m = \left.\frac{\partial I_D}{\partial V_{GS}}\right|_{\text{bias}} \qquad (2.14)$$

b) Substrate transconductance g_{mb}:

$$g_{mb} = \left.\frac{\partial I_D}{\partial V_{BS}}\right|_{\text{bias}} \qquad (2.15)$$

c) Source-drain conductance g_d:

$$g_d = \left.\frac{\partial I_D}{\partial V_{DS}}\right|_{\text{bias}} \qquad (2.16)$$

All the above quantities are evaluated around the quiescent point of the transistor as shown in Fig. 2.6. Each conductance is evaluated by varying the corresponding voltage by a small quantity while keeping the other two constant. If all three voltages are varied simultaneously, the corresponding total variation in the drain current is given by Eq. (2.17):

$$\Delta I_D \simeq \left.\frac{\partial I_D}{\partial V_{GS}}\right|_{\text{bias}} \times \Delta V_{GS} + \left.\frac{\partial I_D}{\partial V_{BS}}\right|_{\text{bias}} \times \Delta V_{BS} + \left.\frac{\partial I_D}{\partial V_{DS}}\right|_{\text{bias}} \times \Delta V_{DS} \qquad (2.17)$$

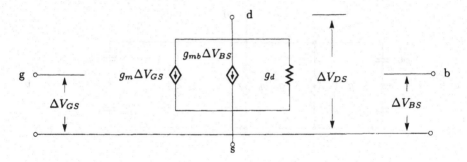

Figure 2.7: MOS transistor small-signal equivalent circuit for low frequencies of operation

$$\Delta I_D \simeq g_m \Delta V_{GS} + g_{mb} \Delta V_{BS} + g_{ds} \Delta V_{DS} \qquad (2.18)$$

Considering zero values for the gate (I_G) and substrate (I_B) currents, Eq. (2.18) provides the small-signal equivalent circuit of the MOS transistor at low frequencies. The circuit schematic is shown in Fig. 2.7.

Taking into account the previous definitions, the MOS transistor (trans) conductances in strong inversion operation, are given by the following equations [4]:

A) Non-saturation region

$$g_m = \frac{1}{2}\mu C'_{ox}\frac{W}{L}V_{DS} \qquad (2.19)$$

$$g_{mb} = \frac{\gamma}{2\sqrt{V_{SB} + \phi_0 + 0.4V_{DS}}}g_m \qquad (2.20)$$

$$g_d = \frac{1}{2}\mu C'_{ox}\frac{W}{L}(V_{GS} - V_T - (1 + \delta)V_{DS}) \qquad (2.21)$$

B) Saturation region

$$g_m = \sqrt{\frac{\mu C'_{ox}}{1 + \delta}\frac{W}{L}I_D} \qquad (2.22)$$

$$g_{mb} = \frac{\gamma}{2\sqrt{V_{SB} + \phi_0 + 0.4V'_{DS}}} \qquad (2.23)$$

$$g_d = \frac{I_D}{V_A} \qquad (2.24)$$

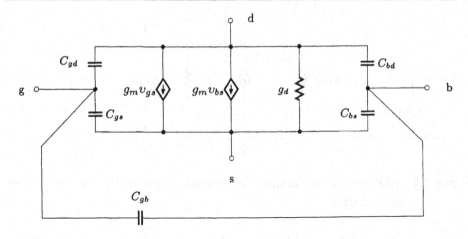

Figure 2.8: Small-signal equivalent circuit at medium frequencies

2.3.2 Small-Signal Equivalent Circuit at Medium Frequencies

A weak variation in the voltage between two of the transistor terminals, causes a corresponding variation in the charge quantities inside the device that has to be compensated externally. Thus, a charge quantity has to be moved from/to a specific transistor terminal. If this variation takes place in short time intervals (i.e. at a high frequency rate), then the derivative $\frac{dq}{dt}$ can take a non negligible value. As long as the charge variation follows the voltage variation in time, the situation is called quasi-static operation: the charge variation Δq is proportional to voltage variation Δv. To the limit:

$$\frac{dq}{dt} = \frac{dq}{dv}\frac{dv}{dt} = C\frac{dv}{dt} \tag{2.25}$$

thus, this variation represents a capacitive current. Such capacitive currents exist between all possible terminal combinations, on top of the conductances presented in Paragraph 2.3.1. The corresponding equivalent circuit is shown in Fig. 2.8.

For a thorough presentation of the influence of the above mentioned capacitances to the operation of the MOS device, the transistor will be split in two parts: the intrinsic part that is responsible for the operation of the device and the extrinsic part that is responsible for all parasitic elements. The intrinsic part of the transistor is the area between source and drain that encompasses the gate, the thin oxide layer and the inversion layer. The extrinsic part of the transistor is divided in different areas each one of which is responsible for a particular parasitic capacitive element of the device. In Fig. 2.9(a), the

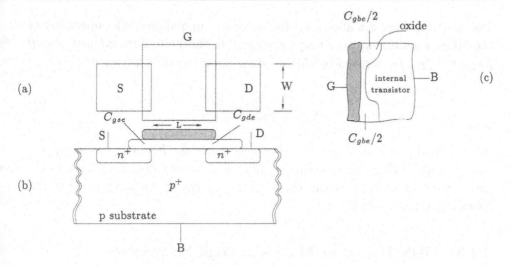

Figure 2.9: Parasitic elements of the nMOS transistor

layout of a typical nMOS transistor is shown. In Fig. 2.9(b) the cross-section is presented while in Fig. 2.9(c), its side view appears [4].

As can be seen in Fig. 2.9, the gate overlaps the transistor's channel towards the source and drain areas. Moreover, in the extrinsic part of the device (regions outside the channel), the oxide layer becomes thick (approximately 40 times thicker than in the area above the channel). This is shown in Fig. 2.9(c). Due to this overlap, two more parasitic capacitances are formed: a capacitance between gate and source (C_{gse}) and a capacitance between gate and drain (C_{gde}) as shown in Fig. 2.9(b). It is obvious that the value of these capacitances is proportional to the channel width W. An additional extrinsic capacitance is formed by the gate and the substrate of the transistor (C_{gbe}) with the thick oxide layer in between, as shown in Fig. 2.9(c). The value of C_{gbe} is proportional to the channel length L. Apart from the above mentioned capacitances, two more extrinsic junction capacitances are formed between substrate and source (C_{bse}) and substrate and drain (C_{bde}). As shown in Fig. 2.9, these capacitances comprise two parts: the bottom area capacitance and the sidewall capacitance. The intrinsic capacitances of the MOS transistor are due to the carrier concentration in the channel area. These capacitances are: C_{gsi}, C_{bsi}, C_{gdi} and, C_{gbi}. Their values depend on the mode of operation of the device (triode or saturation). In both cases, the small-signal capacitances of the model of Fig. 2.8, are given by the following equation:

$$C_{xy} = C_{xyi} + C_{xye} \qquad (2.26)$$

The model of Fig. 2.8 allows for the accurate simulation of the operation of the MOS transistor up to a frequency equal to the 1/10 of its *intrinsic cutoff frequency* f_{T_i}. f_{T_i} is given by the following approximate expression:

$$f_{T_i} \simeq \frac{3}{2} \frac{\mu(V_{GS} - V_T)}{2\pi(1+\delta)L^2} \tag{2.27}$$

The internal cutoff frequency of the transistor is the frequency at which the current gain of the device becomes unity. The current gain is defined as the ratio of the magnitude of the small signal current to the magnitude of the small signal gate voltage.

2.3.3   MOS Transistor Models at High Frequencies

The transcapacitances model

The model presented in the previous paragraph comprise five capacitances between the transistor terminals. The reliability of such a model is extended to a certain frequency of operation. It is evident that these five capacitances are not adequate enough to accurately describe the transistor behavior between all terminal pairs. For this to happen, it is imperative to take into account all possible combinations between transistor terminals and thus to evolve a full quasi-static model. This particular model is called transcapacitances model and its usage results in an increased frequency span validity as opposed to its simple quasi-static counterpart. Particularly, the transcapacitances model is valid up to 1/3 of f_{T_i}. The transcapacitances are defined as follows:

$$C_{kk} = + \left.\frac{\partial q_k}{\partial v_k}\right|_0 \tag{2.28}$$

and

$$C_{kl} = - \left.\frac{\partial q_k}{\partial v_l}\right| \qquad l \neq k \tag{2.29}$$

The negative sign in the above equation results from the fact that Eqs. (2.28-2.29) do not describe the operation of real capacitances but rather, they define model transcapacitances. The transcapacitances model has already been included in the BSIMIII version 3 SPICE MOS transistor model, however without complete success so far.

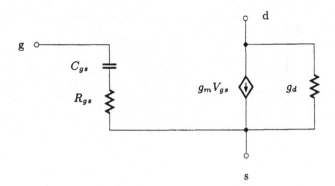

Figure 2.10: First order nonquasi-static model

The first order nonquasi-static model

When the frequency of the signal that is applied to the transistors is very high, the "inertia" of the inversion layer is non negligible anymore and thus there is a certain amount of delay between the cause and the result. For example, the variation of the charge in the gate of the transistor exhibits a phase lag with respect to the voltage signal variation that causes it. Thus, it is evident that nonquasi-static models must be introduced in order to predict the above phenomenon. These models can be used to describe the phase difference between the drain current and the voltage V_{gs}. In a first order approximation, the transconductance g_m of the transistor is replaced by a *transadmittance* y_m:

$$y_m = \frac{g_m}{1 + j\frac{f}{f_x}} \qquad (2.30)$$

where $f_x = 2.5 f_{T_i}$.

The corresponding small-signal equivalent circuit of the intrinsic part of the MOS transistor operating in strong inversion, saturation, is shown in Fig. 2.10. The resistor R_{gs} is not a real ohmic resistor but rather, it represents the inertia of the inversion layer mentioned before. The formulation is $C_{gs} = \frac{2}{3} W L C'_{ox}$ and $R_{gs} = \frac{1}{5g_m}$ The validity of the first order nonquasi-static model is extended up to the f_{T_i} frequency.

Note: The symbol V_{gs} for the definition of the gate-source voltage is used here to stress out that in the nonquasi-static analysis, for all voltage and current quantities, phasors are used instead of the small-signal magnitude quantities employed in the simple quasi-static models.

It is worth noting here that no SPICE-like program is capable of warning the user on the validity of the frequency of operation for the MOS transistor

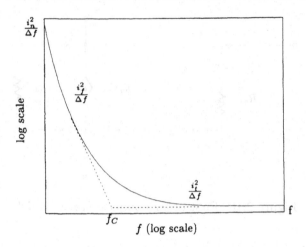

Figure 2.11: MOS transistor noise spectral density

models. This fact must make the RF designer extremely cautious prior to employing any SPICE MOS transistor model in his design. A good technique is to evaluate the f_{T_i} frequency for each device along with the available SPICE model in order to detect the proper frequency limits of operation. This way, erroneous simulations are avoided and the chance of design failures is reduced.

If the frequency of operation of a particular circuit employing MOS transistors is extended beyond the valid frequency range of the existing models, the following technique can be applied: Since the f_{T_i} frequency is inversely proportional to the square of the channel length L, then the transistor can be broken down in smaller elements (using smaller L values) connected in series. This way, the valid frequency range for the model is increased respectively and can be adequate enough for the simulations. However, care must be taken so that the new devices with artificially smaller channel lengths do not exhibit short channel effects.

2.4   NOISE IN THE MOS TRANSISTOR

The current noise spectral density in a MOS transistor operating in strong inversion, is shown in Fig. 2.11. It is noted that in the above figure, both axes are in logarithmic scale. It is obvious from Fig. 2.11 that two distinct regions with respect to noise behavior exist: At low frequencies and up to the corner frequency f_c, the flicker noise is dominant. Above the corner frequency, the thermal noise prevails. Note that the two noise components are uncorrelated.

The thermal noise f_c of a MOS transistor can be modeled by an equivalent ohmic resistor R_n which exhibits the same noise spectral density ($4kTR_n$) as the transistor. The value of this resistor is given by Equation (2.31):

$$R_n = \frac{2}{3} \frac{(1+\delta)^2}{(W/L)\mu C'_{ox}(V_{GS} - V_T)} \frac{1 + \alpha + \alpha^2}{(1 - \alpha^2)(1 - \alpha)} \tag{2.31}$$

An intrinsic problem of the above equations is that for $\alpha = 1$ ($V_{DS} = 0$), the value of R_n is infinite. Due to this fact, many SPICE noise models fail to predict the noise of the MOS transistor in the triode region of operation. The flicker noise spectral density is given by:

$$\frac{\overline{i_f^2}}{\Delta f} = \frac{M g_m^2}{C'_{ox}{}^\alpha W L} \frac{1}{f} \tag{2.32}$$

where α takes values in between 1 and 2 and M is dependent both on the value of α and the transistor structure. Therefore, for high frequency operation, the dominant noise source is the thermal noise generated in the transistor channel. This noise is white noise having the following spectral density [7]:

$$\frac{\overline{i_t^2}}{\Delta f} = 4kT\gamma g_{d0} \tag{2.33}$$

g_{d0} is the drain conductance under zero bias conditions and γ is a bias-dependent coefficient taking values between $\frac{2}{3}$ and 1. For short-channel devices however, (e.g. $L = 0.7\mu m$) γ can take values between 2 and 3. It is therefore obvious that the continuous demand for smaller transistor technologies in RF applications faces a serious obstacle in terms of noise. This excess noise is due to the existence of hot electrons in the transistor channel: High electric fields in short-channel devices increase the electron temperature above the lattice temperature.

An additional noise source comes from the distributed gate resistance [8]. This noise source is modeled by an ohmic resistor connected in series to the thermal noise source at the gate terminal. The value of this resistor R_g and therefore, the corresponding noise, can be drastically reduced if the transistor is laid out in an interdigitized form. More specifically, the value of R_g is given by the following equation:

$$R_g = \frac{R_\square W}{3n^2 L} \tag{2.34}$$

where $R_\square$ is the sheet resistance of the polysilicon gate material, W is the gate width, L is the gate length and n is the total number of gate segments.

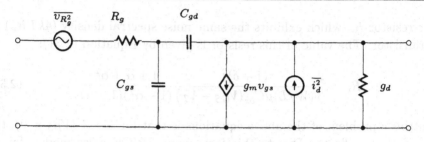

Figure 2.12: MOS transistor noise model

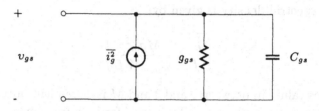

Figure 2.13: MOS transistor gate model

This way, the influence of the gate thermal noise to the total transistor noise becomes negligible. The complete noise model of the MOS transistor is shown in Fig. 2.12.

New effects take place in the noise behavior of the MOS transistor at very high frequencies of operation. More particularly, charge variations in the inversion layer, induce current to the gate of the transistor due to capacitive coupling. As mentioned in the previous section, the phase difference that is caused by the high frequency of operation, is modeled by a real noiseless resistor r_{gs} which is different than the ohmic resistor of the polysilicon layer. The two phenomena combine in one model as shown in Fig. 2.13.

The noise source shown is the current noise source induced to the gate. $\overline{i_g}^2$ and g_{gs} are given by the following formulas [9]:

$$\frac{\overline{i_g}^2}{\Delta f} = 4kT\delta g_{gs} \tag{2.35}$$

and

$$g_{gs} = \frac{\omega^2 C_{gs}^2}{5g_{d0}} \tag{2.36}$$

The above equations hold for the operation of the MOS transistor in the saturation region. The value of δ is $4/3$ for a long channel device. From Eq.

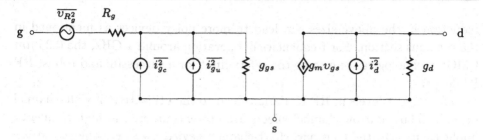

Figure 2.14: Complete noise model of the MOS transistor

(2.36) it is obvious that g_{gs} is proportional to ω^2. Thus, the noise at the gate of the transistor is not white anymore but it is rather considered as "blue". A final comment concerns the correlation of the gate noise to the drain noise. The gate noise can be split into two components: One component uncorrelated to the drain noise and one component correlated to the drain noise, as shown in Eq. (2.37):

$$\frac{\overline{i_g^2}}{\Delta f} = \underbrace{4kT\delta g_{gs}(1 - |c|^2)}_{\text{Uncorrelated}} + \underbrace{4kT\delta g_{gs}|c|^2}_{\text{Correlated}} \tag{2.37}$$

where

$$c \simeq 0.395j$$

The complete noise model of the MOS transistor is shown in Fig. 2.14. A complete noise model for the MOS device is particularly useful in RF IC design wherever a low noise design (i.e. LNA design) is demanded.

2.5 MOS TRANSISTOR TECHNOLOGY LIMITS IN RF IC DESIGN APPLICATIONS

The most important parameters of the MOS device in RF IC applications are the $\mu C'_{ox}$ product and the parasitic capacitances [10]. The $\mu C'_{ox}$ product defines the current needed for the operation of the transistor at a particular frequency while the parasitic capacitances, apart from their influence in the speed of operation of the transistor, also affect the isolation level between the input(s) and the output(s) of the RF circuit due to the parasitic signal paths that are formed.

The major targets in a technology are the reduction of the minimum channel length in order to achieve higher levels of integration as well as the reduction of the oxide thickness to increase the transconductance. However, the reduction in the above sizes can lead to more noisy devices as mentioned in the previous section. For frequencies of operation around 1 GHz, the 0.25 μm CMOS technologies seem to be the safety limit for a successful and robust RF IC design.

The current levels in RF IC designs lead to the utilization of wide-channel devices. Thus, narrow-channel effects are not encountered in high frequency applications. On the contrary, short-channel devices can be employed in the design and therefore, short-channel effects might have to be taken into account. It is thus imperative for the designer to properly select the appropriate transistor model for his simulations. If the channel length in a particular MOS technology becomes very small, there is always a danger that the depletion regions of the source and drain respectively contact each other and the transistor channel is destroyed (punch-through). Therefore, in order to decrease L in a particular technology, the depletion regions have to be reduced. This can be achieved by increasing the doping of the substrate as well as by reducing the inverse bias. Reducing the inverse bias demands the reduction in the supply voltages which is desirable since it can lead to lower power consumption from the circuit provided that it does not affect the dynamic range of the system. However, the increase in doping concentration of the substrate affects the quality of the integrated inductors as will be shown in the next chapter.

2.6   EVALUATION OF EXISTING MOS TRANSISTOR SPICE MODELS

The existence of appropriate SPICE models for high frequency analog integrated circuits is a prerequisite towards a successful and robust design. This is especially true for the MOS transistor where even today (1998), there is no single general-purpose reliable model that covers all areas of operation.

Certain benchmark tests have been derived for the evaluation of the MOS transistor models [11]. Next, the results of such tests on two modern SPICE models are presented. More particularly, the BSIMIIIv2 and BSIMIIIv3 on a 0.5 μm CMOS process are presented. By no means the following presentation can be considered complete; many checks are needed in order to certify the reliability and integrity of a model under all conditions and all regions of operations. For the RF IC design needs, it is indispensable to employ a model accurate for both low voltage and high frequency operation.

Certain general-purpose evaluation criteria for the reliability of the MOS transistor models follow ([3], [11]):

1. Accuracy in the prediction of the small-signal values (g_m, g_{mb}, g_{ds}, ca-

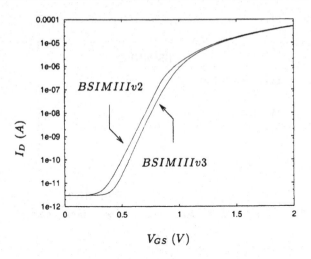

Figure 2.15: $\log I_D$ vs. V_{GS}

pacitances). Function continuity in all regions of operation.

2. Accurate prediction even under nonquasi-static operation of the device.

3. Accurate prediction of the noise performance (both flicker and white) even in the triode region of operation.

4. The above conditions (1-3) must hold under all bias conditions as well as in weak, moderate and strong inversion.

5. The above conditions (1-4) must hold in the complete temperature range of operation.

6. The above conditions (1-5) must hold for all valid W and L values of the transistor channel.

2.6.1 Simulation Results

Two different SPICE-like simulators have been used for the evaluation of BSIMIII models namely ELDO and HSPICE. No significant difference in the results were reported. In all cases, a nMOS transistor with $W = 10\mu m$ and $L = 10\mu m$ was used unless mentioned differently.

a) Drain current in the weak-moderate-strong inversion regions (I_D). In Fig. 2.15, the drain current I_D (in log scale) as a function of V_{GS} for constant V_{DS} and V_{SB} values is shown. It is obvious that the transition regions from weak to moderate and from moderate to strong inversion are quite abrupt in the case of the BSIMIIIv2 model.

b) Transconductance to drain current ratio (g_m/I_D) In Fig. 2.16 the ratio of transconductance to drain current as a function of the drain current (in log

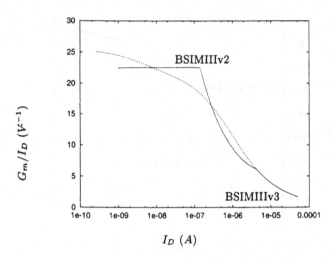

Figure 2.16: g_m/I_D vs. I_D for BSIMIIIv2 and BSIMIIIv3 models

scale) is shown for both the BSIMIIIv2 and the BSIMIIIv3 model.

It is obvious that in the BSIMIIIv3 model, all the abrupt transitions present in the BSIMIIIv2 model have been eliminated.

c) Substrate transconductance (g_{mb}). The substrate transconductance is a very important parameter in an analog design if the body effect is present. In Fig. 2.17, the g_{mb}/I_D to V_{BS} ratio for a constant V_{GS} value is shown. In this case, both models fail to predict the actual behavior of the device.

d) Drain-source conductance (g_{ds}). In Fig. 2.18, the drain-source conductance as a function of V_{DS} and for $V_{GS} = 2V$ is shown. The transition from the triode region to saturation is smooth in the BSIMIIIv3 model and abrupt in the BSIMIIIv2 model.

e) Drain current (I_D) versus V_{DS}. In Fig. 2.19, the drain current versus V_{DS} for both models is shown. It is evident that the two models predict different current values. Comparison to experimental results will prove the accuracy of each model in every case.

f) Gain coefficient (g_m/g_{ds}). In Fig. 2.20, the g_m/g_{ds} ratio versus V_{GS} (in log scale) for both models is shown. The transition between the various regions of operation is quite smooth in the BSIMIIIv3 case.

g) Thermal noise. The prediction of the thermal noise behavior of the MOS transistor while operating in the triode region, is a test that nearly all SPICE-like models fail to pass. In this case, the transistor is equivalent to an ohmic resistor with value $R = 1/g_{ds}$ and, therefore, SPICE must predict the corresponding noise level. Unfortunately, the BSIMIIIv2 completely fails

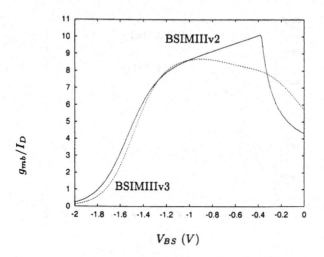

Figure 2.17: g_{mb}/I_D vs. V_{BS}

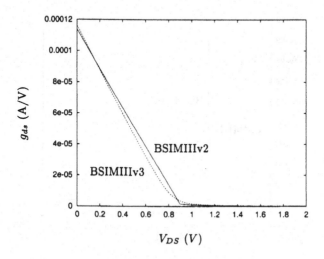

Figure 2.18: g_{ds} vs. V_{DS}

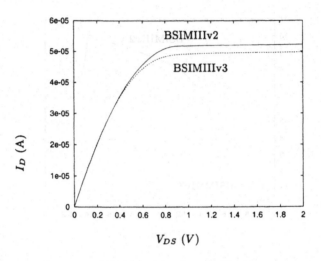

Figure 2.19: I_D vs. V_{DS}

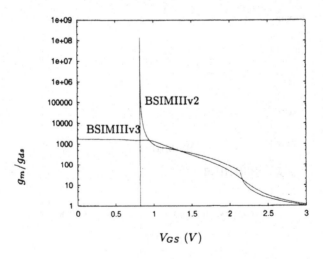

Figure 2.20: g_m/g_{ds} vs. V_{GS}

in our case while the BSIMIIIv3 model gives correct results under certain conditions of operation. In every case, only the comparison of the predicted value from the model to the measured one can prove the reliability of the model.

h) The transconductance at high frequencies. This is a very useful test for the operation of the transistor at high frequencies. For this purpose, a quite big transistor $(L = 100\mu m)$ was employed in simulations. This transistor is biased in the strong inversion, saturation region. Prior to performing the simulations, all the parasitic parameters of the model are zeroed (e.g. junction and overlap capacitances, resistances etc.) Usually, SPICE models predict a constant function of the transconductance versus frequency - a result that is wrong; the drain current is not frequency independent. In order to reveal this frequency dependence, the transistor is split into more than one devices connect in series so that the sum of their channel lengths equals that of the original device. Thus, based on Eq. (2.27) and the discussion in Paragraph 2.3.3, the accuracy of the MOS model is extended to higher frequencies and the simulation results are realistic. In Fig. 2.21, the ac response of I_D for the BSIMIIIv2 and BSIMIIIv3 models is shown. In the BSIMIIIv2 case, the transistor with channel length of 100 μm has been replaced by two equal series-connected devices with $L = 50\mu m$ each. It is noted that the BSIMIIIv3 model completely fails. This is due to the improper implementation of the transcapacitances principle in this model. The most appropriate model in this case would be a nonquasi-static model. For the time being (1998), the only nonquasi-static model employed in SPICE-like simulators is the EKV model.

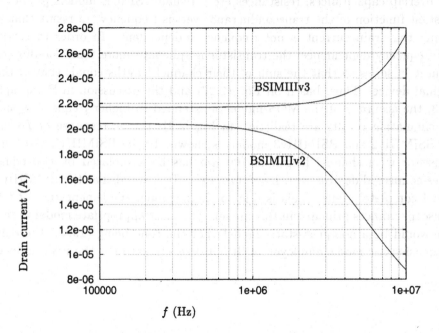

Figure 2.21: drain current ac response for the BSIMIIIv2 and the BSIMIIIv3 models

Bibliography

[1] R.H. Yan, K.F. Lee, D.Y. Jeon, Y.O. Kim, et al., "High performance 0.1 micron room temperature Si MOSFETS," in *Digest of Technical Papers, 1992 Symposium on VLSI Technology*, Seattle, WA, 2-4 June 1992, pp. 763–766.

[2] C. Jian, S. Parke, J. King, F. Assaderaghi, et al., "A high speed SOI technology with 12ps/18ps gate delay operating at 1.5V," in *Proceedings of IEEE Electron Devices Meeting*, San Francisco, CA, 13-16 Dec. 1992.

[3] Y.P. Tsividis, *Operation and Modeling at the MOS Transistor*, Mc Graw-Hill, 1987.

[4] Y.P. Tsividis, *Mixed Analog-Digital VLSI Devices and Technology*, Mc Graw-Hill, 1995.

[5] P.R. Gray and R.G. Meyer, *Analysis and Design of Analog Integrated Circuits*, John Wiley, 1992.

[6] A.S. Machado Ed., "Low Power HF Microelectronics," *IEE Circuits and Systems*, 1996.

[7] D.K. Shaeffer and T.H. Lee, "A 1.5V, 1.5 GHz CMOS Low Noise Amplifier," *IEEE JSSC*, vol. 32, no. 5, pp. 745–759, May 1997.

[8] R.P. Jindal, "Noise associated with distributed resistance of MOSFET gate structures in integrated circuits," *IEEE Trans. Electron Devices*, vol. ED-31, pp. 1505–1509, Oct. 1984.

[9] A. van der Ziel, *Noise in Solid State Devices and Circuits*, Wiley, New York, 1986.

[10] Q. Huang, F. Piazza, P. Orsatti, and T. Ohguro, "The Impact of Scaling Down to Deep Submicron on CMOS RF Circuits," in *Proceedings of the*

23rd ESSCIRC Conference, Southampton UK, 16-18 September 1997, pp. 132–135.

[11] Y.P. Tsividis and K. Suyama, "MOSFET Modeling for Analog Circuit CAD: Problems and Prospects," *IEEE JSSC*, vol. 29, no. 3, pp. 210–216, March 1994.

CHAPTER
3

INTEGRATED PASSIVE ELEMENTS AND THEIR USAGE AT HIGH FREQUENCIES

3.1 INTRODUCTION

For the design of integrated circuits in the radio frequency band, equally important to active elements (transistors) are also the passive ones (inductors, capacitors, resistors). For example, the existence of high quality capacitors and inductors significantly determines the circuit performance and usually their shortage in an IC technology prevents the design of high frequency systems. Capacitors and ohmic resistors are elements that can be easily fabricated in analog silicon technologies. Recently, the fabrication of capacitors has been extended to digital (CMOS) technologies [1,2] due to their low cost and high level of integration they exhibit. Integrated inductors had not been fabricated within silicon processes until recently [3,4]. The main reason was the lack of a satisfactory model of the electrical and magnetic performance of the element. Recently, (1997) a complete model for integrated inductors over silicon substrates, along with a CAD tool, have been presented [5,6]. This effort will significantly boost the usage of inductors in silicon integrated circuits, something that has been very common and for many years in *GaAs* technologies.

The fabrication of variable integrated capacitors (varactors) is of great importance, lately, due to their wide usage in the fabrication of integrated oscillators (VCO).

Finally, parasitic phenomena that are induced by the silicon substrate and mainly by the package of the integrated circuit have a critical part in the design of integrated circuits for RF applications. For this reason, the accurate modeling and the inclusion of an electrical equivalent in the designed system is absolutely necessary.

3.2   INTEGRATED INDUCTORS OVER Si SUBSTRATES

Modern telecommunication systems increasingly demand the reduction of power consumption and production cost along with the increase of integration density. The broad usage of inductors in integrated circuits can lead to the design of improved systems, with lower noise floor and higher levels of integration. Towards this direction, great effort has been put on the accurate modeling of the performance of integrated inductors on silicon substrates ([4-7]). The existence of a generic lumped element model of the performance of the integrated inductor in any silicon technology and in a wide frequency band, is a very valuable tool for the $RFIC$ designer. Among other things, it will give the opportunity to optimize the performance of integrated inductors according to the requirements of every application, avoiding the need of an initial fabrication cycle and measurement of test inductor structures. By using scalable lumped element models, production cost and development time of integrated circuits are significantly reduced.

In literature, several topologies of integrated inductors have been presented mostly dominated by square and octagonal structures. In addition, several attempts have been made to fabricate integrated inductors with more than one layers taking advantage of the numerous metalization layers that are usually provided by modern silicon technologies.

The basic idea of the model of an integrated inductor, which will be presented later on, is the generation of two-port network for every segment of the spiral inductor structure.

3.2.1 Electrical Model of the Performance of the Integrated Inductor

Fig. 3.1 presents a typical structure of two metallic microstrips in a silicon technology. Underneath the metallic material of the microstrips, there is insulating material (usually a thick layer of SiO_2). The microstrips are usually fabricated on the upper available metal layer of the technology, in order to minimize, as much as possible, the parasitic capacitances to the substrate. The microstrips can also be placed over a well which resides in the silicon substrate, if this is necessary. From Fig. 3.1 it is obvious that there is a mutual electromagnetic coupling between the two microstrips and also between every microstrip and the lossy silicon substrate. A complete model that describes the electrical and magnetic behavior of the integrated inductor would aim to calculate accurately, apart from the overall inductance and ohmic losses of the microstrips that comprise the spiral inductor, all the aforementioned parasitic coupling and crosstalk.

The following geometrical quantities are defined in Fig. 3.1: $l=$ length of the microstrip, $W=$ width of the microstrip, $t=$ thickness of the microstrip, $s=$ distance between two adjacent parallel microstrips, $d=$ distance between the middle of the cross-section of two adjacent and parallel microstrips, $h_{SiO_2}=$ thickness of the insulating material SiO_2, $h_{Si}=$ thickness of the substrate.

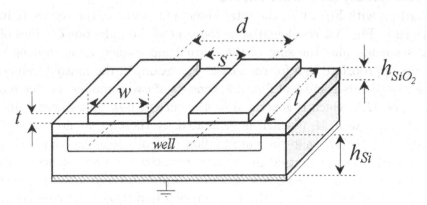

Figure 3.1: Structure of two microstrips on Si substrate

b) Calculation of the inductance of the integrated inductor

Fig. 3.2 shows an overview of the geometry of square integrated inductor. The inductance of a simple segment (microstrip) is given by Eq. (3.1) [8,9]:

$$L = 0.002l[\ln(2l/(W+t)] + 0.50049 + [(W+t)/3l] \qquad (3.1)$$

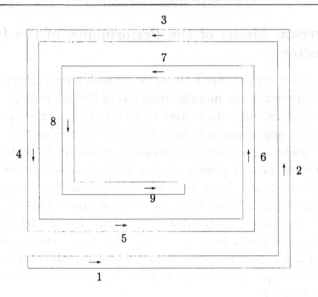

Figure 3.2: Cross-section of square inductor

where L is the inductance in μH, l is the length of the conductor in cm, W and t are the dimensions of the rectangular cross-section of the microstrip as they have already been defined above.

Starting with Eq. (3.1), the total inductance value of the square inductor displayed in Fig. 3.2 is calculated as the sum of the inductance values of the spiral segments plus the sum of the mutual inductance values among them with the appropriate sign for each case. For example, the mutual inductance between segments 1 and 5 of Fig. 3.2, consists of two components: the mutual inductance $M_{1,5}$ which is generated by the current flow in segment 1 and the mutual inductance $M_{5,1}$ which is generated by the current flow in segment 5. Due to the fact that the current flow in both segments has the same direction, the overall mutual inductance between the two segments will be $+(M_{1,5}+M_{5,1})$. On the contrary, due to the opposite direction of the currents flowing in segments 1 and 7, the total mutual inductance between them will have a negative sign. As a result, the total inductance value of the inductor is given by the following relation:

$$L_{tot} = \sum L_{seg} + \sum M \qquad (3.2)$$

where L_{seg} is the inductance value of every segment and $\sum M$ is the algebraic sum of the mutual inductance values of the all the segments calculated in pairs.

The mutual inductance between two parallel straight segments is given by

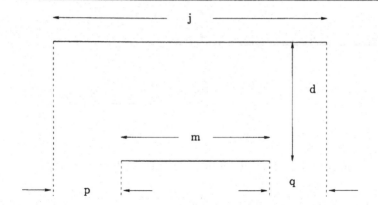

Figure 3.3: Two straight conductor segments

Eq. (3.3):

$$M = 2lU \tag{3.3}$$

where l is the length of the two conductors and U is a factor that is given by Eq. (3.4):

$$U = \ln\left\{(l/GMD) + [1 + (l^2/GMD^2)]^{\frac{1}{2}}\right\} - [1 + (GMD^2/l^2)]^{\frac{1}{2}} + (GMD/l) \tag{3.4}$$

where GMD is the geometrical mean distance between two segments and which is approximately equal to the distance d between the midpoints of the two segments. The exact expression for the GMD is given by the following series:

$$\ln GMD = \ln d - \left\{\frac{1}{12}\left(\frac{d}{W}\right)^2 + \frac{1}{60}\left(\frac{d}{W}\right)^4 + \frac{1}{168}\left(\frac{d}{W}\right)^6 + \frac{1}{360}\left(\frac{d}{W}\right)^8 + \cdots\right\} \tag{3.5}$$

Fig. 3.3 exhibits the geometrical layout of two straight conductor segments with length of j and m respectively. The geometric mean distance between them is GMD. Considering that $M_{j,m} = M_{m,j}$ then:

$$2M_{j,m} = +(M_{m+p} + M_{m+q}) - (M_p + M_q) \tag{3.6}$$

where

$$M_{m+p} = 2l_{m+p}U_{m+p} = 2(m+p)U_{m+p} \tag{3.7}$$

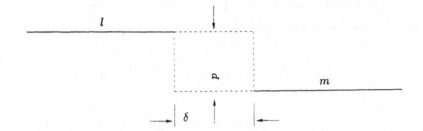

Figure 3.4: Two straight segments without overlap

Also for $p = q$ then it stands:

$$M_{j,m} = M_{m+p} - M_p \tag{3.8}$$

and for $p = 0$ it stands :

$$2M_{j,m} = (M_j + M_m) - M_q \tag{3.9}$$

To cover all types of the layout design of integrated inductors (octagonal and in general polygonal and 3-D), all possible cases of the relative placement of two straight segments in space must be considered.

Fig. 3.4 shows two straight conductors placed parallel to each other but with no overlap. The distance δ is the distance between the endpoint of the first segment and the projection of the endpoint of the second segment on the first. If there is an overlap of the projection of the second segment to the first then δ becomes negative. The mutual inductance of the two segments is given by the following relation:

$$2M = (M_{l+m+\delta} + M_\delta) - (M_{l+\delta} + M_{m+\delta}) \tag{3.10}$$

Fig. 3.5 displays all the possible cases of straight segments that are at an angle ϕ. In the case where the two segments are placed as shown in Fig. 3.5 (a), then their mutual inductance is given by Eq. (3.11):

$$M_{l,m} = 2\cos\phi\left[l\tan h^{-1}\left(\frac{m}{l+y}\right) + m\tan h^{-1}\left(\frac{l}{m+y}\right)\right] \tag{3.11}$$

In Fig. 3.5 (b), the intercept point P is placed outside the two segments and their mutual inductance is given by Eq. (3.12):

$$M_{l,m} = 2\cos\phi[M_{\mu+l,\nu+m} + M_{\mu\nu} - (M_{\mu+l,\nu} + M_{\nu+m,\mu}) - \Omega d_z/\sin\phi] \tag{3.12}$$

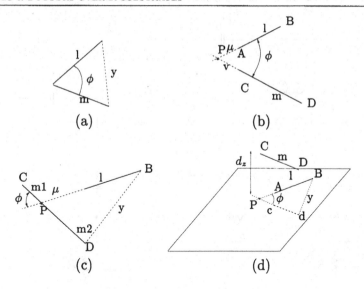

Figure 3.5: Two straight segments at an angle

where

$$\Omega = \frac{\pi}{2} + \tan^{-1}\left[\frac{d_z{}^2\cos\phi + lm\sin^2\phi}{d_z y \sin\phi}\right] - \tan^{-1}\left(\frac{d_z \cos\phi}{l\sin\phi}\right) - \tan^{-1}\left(\frac{d_z \cos\phi}{m\sin\phi}\right) \tag{3.13}$$

Eq. (3.12) is generic and refers to cases in which the two segments are placed either on the same plane or in space belonging to different planes. Variable Ω is defined by Eq. (3.13) and refers to the case in which the two segments are placed in space ($d_z \neq 0$ - Fig. 3.5 (d)).

The most complex case is presented by Fig. 3.5 (c), where the intercept point P is on one of the segments. The approach for this case is to divide the segment CD into two subsegments: CP and PD. The total mutual inductance is calculated as the sum of the mutual inductances M_1 and M_2 between the segments AB and CP, and AB and PD respectively.

a) Calculation of the parasitic ohmic resistance of the integrated inductor

For every segment of the integrated inductor (Fig. 3.1), its parasitic ohmic resistance is calculated and is connected in series to its inductance. The value of the resistance is given by Eq. (3.14):

$$R = R_{sh}l/W \tag{3.14}$$

where R_{sh} is the sheet resistance of the conductor material that the inductor is made of.

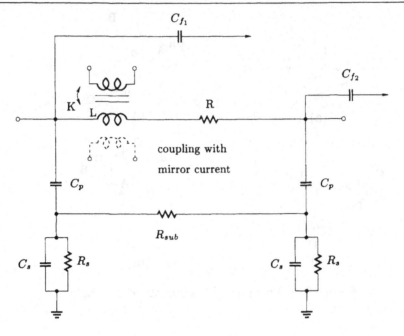

Figure 3.6: Schematic diagram of the SPICE model of a conductive segment
of an integrated inductor

At this point, and since the algorithm that calculates the inductance, mutual inductance and ohmic resistance values has already been explained, the
complete electrical model for every segment of the spiral inductor will be presented [5]. The schematic diagram of the model is shown in Fig. 3.6. The
figure displays the basic lumped elements that describe and model all the phenomena related to every integrated spiral inductor. The inductance of every
segment is divided into two parts: (I) its pure self-inductance L and (ii) its mutual inductance with every other segment of the inductor. This dependence is
described electrically with the respective transformers that are employed with
a coupling coefficient K which is given by Eq. (3.15):

$$K = M_{1,2}/\sqrt{L_1 L_2} \tag{3.15}$$

In Fig. 3.6, the inductor with the dotted lines stands for the magnetic
coupling of the spiral inductor with its mirror element, which is considered
to exist inside the substrate region, since the silicon substrate is a conductive
material but with non-zero finite resistance. The ohmic losses of the conductive
part are represented by resistance R.

c) The parasitic capacitances of the integrated inductor.

The lumped element model is enriched with the parasitic capacitances, as they are presented in Fir. 3.6: The main parasitic capacitance is the one that is formed between the metallic conductor of the segment and the silicon substrate. This capacitance appears in the model as two capacitances C_p that are connected ate the two terminals of the segment. It should be clarified here that this capacitance consists of the plate capacitance between the conductor and the substrate, and also of the fringe capacitance.

Other side capacitances are formed between parallel adjacent segments (see Fig. 3.1), for example between segment 3 and segment 7 of the inductor displayed in Fig. 3.2. This capacitances are C_{f_1} and C_{f_2} shown in Fig. 3.6 for the respective segments of the conductor.

Capacitances C_p of Fig. 3.6 are given by the following equation:

$$C_p = \varepsilon_0 \varepsilon_r W / h \qquad (3.16)$$

Eq. (3.16) refers only to the part of the capacitance between the bottom surface of the conductor and the substrate. There is an equivalent relation that also describes the fringe capacitances. ε_r is the relative dielectric constant of the insulating material that exists between conductor and substrate (SiO_2 in the case of Fig. 3.1).

The calculation of capacitances C_f between adjacent conductor segments is more complicated. The reader is referred to [10] for a description of the equivalent effects.

d) Silicon substrate modeling.

As it is presented in Fig. 3.6, the substrate is modeled by connecting a complex conductance between the earth and the capacitors of the insulating material at both terminals of the conductive segment. This complex conductance consists of a resistance R_s connected in parallel with a capacitor C_s. The conductance G_s is given by the following relation [11]:

$$G_s = \hat{\sigma} \frac{W}{\hat{h}} \qquad (3.17)$$

where

$$\hat{\sigma} = \sigma \left(\frac{1}{2} + \frac{1}{2\sqrt{1 + 10h/W}} \right) \qquad (3.18)$$

and

$$\hat{h} = \frac{W}{2\pi} \log \left(\frac{8h}{W} + \frac{4W}{h} \right) \qquad (3.19)$$

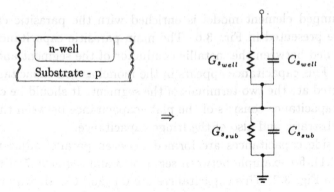

Figure 3.7: Modeling of multiple substrate layers

σ is the conductance of the substrate which performs as a semiconductive material with losses. The capacitance C_s is given by Eq. (3.20) [12]:

$$C_s = \frac{W + \Delta W'}{h} \varepsilon_0 \varepsilon_{r_{si}} \qquad (3.20)$$

where
$$\Delta W' = \left(1 + \frac{1}{\varepsilon_{r_{si}}}\right) \frac{\Delta W}{2}$$
and
$$\Delta W = \frac{t}{\pi} \ln\left[4e/\sqrt{\left(\frac{t}{h}\right)^2 + \left(\frac{1/\pi}{W/t+1.1}\right)^2}\right]$$

Between the two complex conductances, the substrate resistance R_{sub} is connected and is given by:

$$R_{sub} = \rho_{si} l / (W h_{si}) \qquad (3.21)$$

If the technology comprises more than one substrate layers (e.g. wells or buried layers) with different specific resistance, then each one of them is modeled by a complex conductance $C_{si}//G_{si}$ which is connected in series with the equivalent conductance of the next substrate material (Fig. 3.7).

Finally, a very important effect that should be taken into account in the performance modeling of the integrated inductor, is the skin-effect which appears to affect both the inductance and the resistance of the inductor. A complete description exists in [13] and the quantitative calculations that are referred therein have been used for the development of the tool described in [5-6].

3.2.2   Measurements of Integrated Inductors - Comparison with Simulation Results

The existence of a software tool that is able to accurately model the performance of an integrated inductor on silicon substrate would be a valuable aid to the RF IC designer. Spiral Inductor Simulation Program (SISP), a CAD tool which has been developed by the Microelectronic Circuit Design Group of the NTUA [5-6], has been based on the modeling technique presented in the previous paragraph and has been successfully tested against measurements of inductor structures fabricated in various silicon technologies (bipolar, CMOS and BiCMOS).

Nowadays (1997), several microelectronic industries have begun to fabricate inductors in their conventional technologies, by using metalization layers made of aluminum-based alloys. The inductance values that have been achieved are at most a few tens of nH, while their quality factor is between 5 and 7. It is obviously necessary to optimize integrated inductors' performance so that they can be employed in wider range of applications. Towards this direction, existing silicon technologies are currently being modified in order to improve the performance of inductors. For example, the use of other materials for the fabrication of on-chip inductors such as gold ([14]), can significantly increase the value of the quality factor of the inductor. Similar benefits can be achieved by modifying the doping of the silicon substrate. The increase of the specific resistance of the substrate improves the quality factor of the inductor, as well as its resonance frequency which increases, too. In such a case, the semiconductive behavior of the silicon substrate tends to resemble to that of a high resistive substrate such as GaAs.

SISP, is capable on the one hand, to exhibit the improvement of the inductor performance by the appropriate modifications that are made to the silicon technology while on the other hand is capable to show how can someone exploit at most an existing technology. The latter can be achieved by creating useful nomographs, as it will be subsequently shown, which can be exploited by designers so that they can easily choose the most appropriate inductor for each specific application.

In the following paragraphs, integrated inductor measurement results combined with simulation results of SISP models are presented to prove the accuracy of the CAD tool. The measurements of the integrated inductor structures were made with a probe station setup which is used to achieve a direct contact of the measuring instrument with the integrated circuit. The measurements are carried out by using a network analyzer which can measure all four S parameters of a two-port network (in this case, the S parameters of the integrated inductor). From the S parameters of the passive structures, its Y parameters are extracted and are used for the calculation of the inductance value L and quality factor value Q of the integrated inductor based on the following equations:

$$L = \frac{\text{Im}\{1/Y_{11}\}}{2\pi f} \tag{3.22a}$$

$$\text{and} \quad Q = \frac{\text{Im}\{1/Y_{11}\}}{\text{Re}\{1/Y_{11}\}} \tag{3.22b}$$

Both quantities L and Q are plotted afterwards versus frequency and the designer can have as a result a complete overview of the performance of the on-chip inductive element in the frequency bands of interest. It should be emphasized that SISP can also give reliable results by accurately predicting the inductor performance regarding its resonance frequency. Resonance frequency is the frequency where the integrated inductor stops behaving as an inductor and starts behaving as a capacitive element, being practically useless. Consequently, resonance frequency becomes a very critical parameter that should be accurately modeled in order to include an inductor in an integrated circuit.

Fig. 3.8, presents the results of a comparison of the measurement and simulation of a square spiral inductor with three turns, outer dimension of $245.5\mu m$, track width of $W = 12.5\mu m$ and distance between adjacent tracks of $S = 5\mu m$. The inductor has been fabricated in a bipolar silicon technology.

Fig. 3.9, presents the results of a comparison of the measurement and simulation of an octagonal spiral inductor with eight and a half turns, radius of $250\mu m$, track width of $W = 16\mu m$ and distance between adjacent tracks of $S = 8\mu m$. This inductor has also been fabricated in a bipolar silicon technology.

Fig. 3.10, presents the results of the simulation and measurement of the inductance of an inductor which has been designed and fabricated with two metalization layers as it is shown in the figure. For the fabrication of such an inductor, the first and the third layers have been used, of a CMOS technology with three metal layers. The outer dimension of the square double-layer spiral inductor is 250 μm.

From Fig. 3.10, someone may notice that the measurement results have been accurately predicted, especially the frequency region near the self-resonance frequency of the inductor, where it exhibits the highest sensitivity.

Due to this level of accuracy, SISP is a valuable design tool especially in the case of a tuned LC tank that operates as the load of a tuned amplifier:

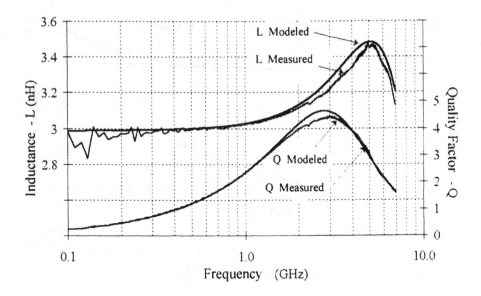

Figure 3.8: Measurement / Simulation results of square spiral inductor

The LC load can be substituted by an integrated inductor which operates near its self-resonance frequency range.

A last remark regarding the inductance value of a double-layer spiral inductor: From Fig. 3.10 it is obvious that the double-layer structure exhibits an inductance value that is even 5 times higher than that of its planar counterpart.

The study of the coupling of two neighboring spiral inductors, one beside the other, on common silicon substrate is also of great interest. This study is very useful for the design of more complex RF systems where the coexistence of integrated inductors on the same silicon die is absolutely necessary. SISP is capable to predict the coupling between these on-chip elements. Fig. 3.12 displays the results of simulation and measurement of the coupling of two square spiral inductors placed one beside the other. The configuration is shown in Fig. 3.13(a).

The same figure exhibits a square double-layer spiral inductor (Fig. 3.13(b)), whose performance has already been presented in Fig. 3.10 and an octagonal spiral inductor (Fig. 3.13(c)), whose performance has also been presented in Fig. 3.9.

The coupling between two adjacent square spiral inductors is measured with parameter S_{21} considering the group of the two inductors as a single two-

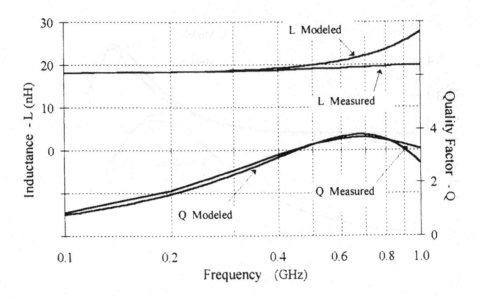

Figure 3.9: Measurement / Simulation results of octagonal spiral inductor

port network.

The capability of SISP in predicting accurately the coupling among all on- chip inductors of the same integrated circuit can be used for the area optimization of the IC in a given design. As an example of a multi-inductor circuit, which is analytically presented in Chapter 5 of this book, the design of an RF CMOS tuned amplifier is presented that employs two integrated inductors. Fig. 5.13 presents the response of the gain of the amplifier for distinct layout schemes of the two spiral inductors. Their relative placement, apparently, significantly affects the performance of the circuit.

Having proven the reliability of SISP with measurements, one can use it to extract valuable results regarding the design of integrated circuits at high frequencies. For example, Fig. 3.14 displays the comparison of simulation results (inductance and quality factor) between various inductor structures. Precisely, three inductors are presented: an octagonal, a square and a double-layer that occupy the same area on the chip. All three inductors have 8.5 turns, $W = 16\mu m$ and $S = 3\mu m$. From the simulation results, it is obvious that the octagonal structure exhibits higher quality factor compared to the other two structure, as well as wider frequency range of operation(range in which the element behaves inductively). On the contrary, the double-layer structure exhibits very higher inductance value but very lower quality factor

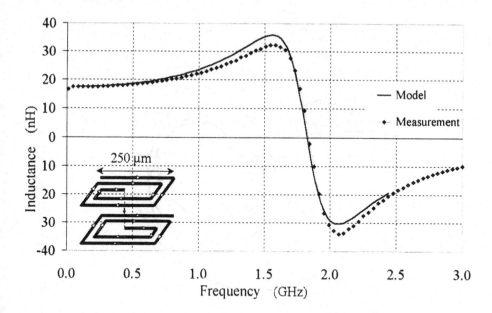

Figure 3.10: Measurement / Simulation results of double-layer spiral inductor

due to the capacitive coupling between the two metal layers that form the structure.

Another handy and useful tool that can be created by using SISP is a nomograph. Nomographs can be used by the designer for the proper selection of the most appropriate inductor for a specific application. For example, Fig. 3.15 shows the variation of the inductance (Fig. 3.15(a)) and of the quality factor (Fig. 3.15(b)) of a family of square inductors, as a function of the number of turns (n) and the track width (W). The results are given at a specific frequency (3GHz in this case) while all the inductors under investigation have a constant outer dimension of $D_0 = 250\mu m$ and distance between tracks of $S = 3\mu m$.

Another example of a very useful nomograph is presented in Fig. 3.16. In this case, a family of octagonal spiral inductors has been selected with a constant radius of $200\mu m$ and their inductance and quality factor are plotted versus frequency and number of turns n. Through these nomographs it is possible to detect higher L and Q values as well as the equivalent self-resonance frequency of octagonal inductors of the same footprint but with varying number of turns.

Finally, SISP can be used for the generation of nomographs that are useful for the evaluation of IC technologies regarding spiral inductors and how they

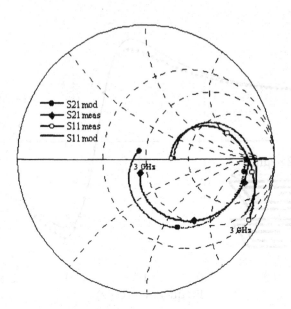

Figure 3.11: Measurement / Simulation results of double-layer spiral inductor:
            S- parameters

affect the performance of inductive passive elements.

Fig. 3.17 displays the dependence of the quality factor on the thickness of the metalization layer that is used for the fabrication of the integrated inductor. The improvement of Q when the metal thickness t increases is obvious and expected. The nomograph in Fig. 3.17 comes from a family of square inductors with 5 turns, outer dimension of $D_0 = 300\mu m$, track width of $W = 14\mu m$ and track spacing of $S = 3\mu m$.

It is worth observing that the resonance frequency of the whole family of inductors has not changed at all, when the metal track thickness increases. In the same way, someone may discover how the variation of other technological parameters can affect the performance of integrated inductors (e.g. the variation of the substrate doping) before realizing any variations that can be very costly and time-consuming.

To conclude the presentation, integrated inductors and their proper modeling is a field of continuous research and their application by the modern microelectronic industry, surely will continue to reveal many results in the near future.

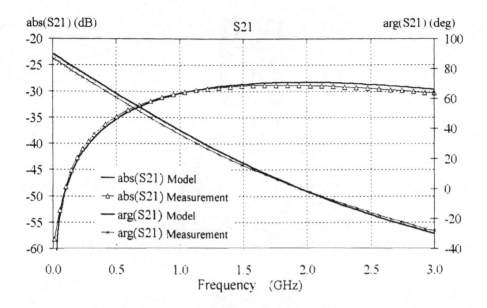

Figure 3.12: Measurement / Simulation results of S_{21} parameter of two adjacent spiral inductors

3.3 VARIABLE CAPACITORS (VARACTORS) IN INTEGRATED TECHNOLOGIES

Varactors in an integrated technology are fabricated using diodes. Fig. 3.18 presents the circuit symbol of a varactor diode. The integration of the one-dimension Poisson equation gives the capacitance value of the varactor [15]:

$$C \equiv \frac{\partial Q_c}{\partial V} = \left[\frac{qB(\varepsilon_s)^{m+1}}{(m+2)(V+V_{bi})}\right]^{1/(m+2)} \sim (V+V_{bi})^{-s} \qquad (3.23)$$

where $s = \frac{1}{m+2}$, Q_c is the load per unit area, ε_s is the dielectric constant of the material and V_{bi} is the junction potential (built-in). In a generic case, the doping of the semiconductor material is given by the following relation:

$$N = Bx^m \qquad (3.24)$$

where parameter x gives the relative position to the junction on the horizontal axis ($x = 0$ to W) and $B = N_B$ when $m = 0$ (uniform material doping). For $m = 1$ the doping varies linearly, but for $m < 0$ the doping of the junction is hyperabrupt. Fig. 3.18(b) displays a simplified schematic diagram of the

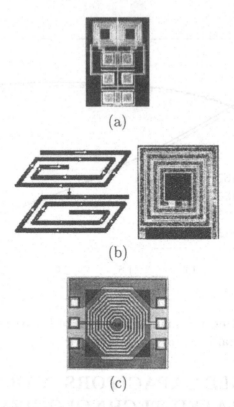

Figure 3.13: Microphotographs of integrated inductors: (a) Two adjacent square inductors, (b) Square double-layer spiral inductor, (c) Octagonal inductor

varactor diode. C_j is the junction capacitance, R_p is the equivalent resistance of the recombination procedure and surface leakage current, and R_s is the series parasitic resistance. The quality of a varactor is given quantitatively by the quality factor Q which is the ratio of the highest value of stored energy to the total energy that is dissipated during a period in the element:

$$Q \simeq \frac{\omega C_j R_p}{1 + \omega^2 {C_j}^2 R_p R_s} \quad (3.25)$$

The quality factor of a varactor is very essential parameter in the design of integrated oscillators (VCOs) because together with quality factor of the integrated inductor is used to define the performance of the circuit and specifically the phase noise. At low frequencies, resistance R_p has an important role in the quality of the varactor, while in high frequencies this role goes to R_s

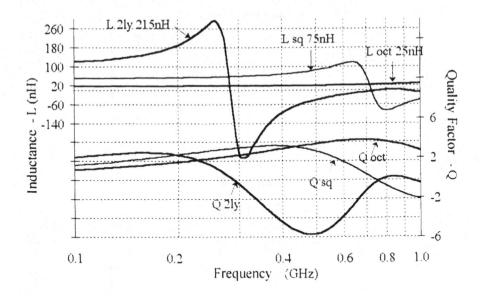

Figure 3.14: Comparison of the performance of various integrated inductor structures

which dominates. As a result, the quality factor of varactors for frequencies higher than a few MHz is given by [16]:

$$Q = \frac{1}{\omega R_s C_j} \tag{3.26}$$

As the inverse bias increases the quality factor increases, too. However, the phenomenon is constrained by the break-down voltage of the junction diode.

In a bipolar technology for RF applications, it is possible to implant a p-type dope in a small depth for the creation of the inner part of the base of an npn transistor. This material combined with the epitaxial layer of n-type are used to form the varactor. In order to minimize the intrinsic parasitic resistances, the buried layer, which is employed for the reduction of the collector resistance of the npn transistor and the junction with the surface of silicon substrate, may also be used. In addition, n-type dopes may also be used in a greater depth, that are connected to the buried layer (sinkers) (see also Fig. 1.2(a)). The base resistance is, in turn, decrease by storing an aluminum layer on its surface. By employing all the above techniques, the resistance of a varactor is decreased, thus its quality factor is increased.

According to the above, a detailed equivalent schematic diagram of a varactor in a bipolar integrated technology is presented in Fig. 3.19. In this figure

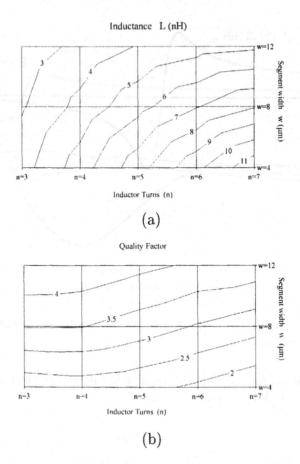

Figure 3.15: Inductance and quality factor of square spiral inductors as a function of the track width W and the number of turns n

C_{sub} and R_{sub} are the parasitic elements of the substrate, C_α is the capacity of the junction contact surface of the two materials (base p-type and epitaxial layer n-type), C_f is the capacitance of the sidewalls, R_α and R_f are the equivalent parasitic resistances and finally R_{sp} and R_{sepi} represent the ohmic losses of the metallic contacts of the two terminals. The junction resistances R_{sp} and R_{sepi} can be reduced if the varactor is formed by many elements connected together in parallel. Using this technique the quality factor of the varactor is increased and may reach theoretically the value of 20 in frequencies around 1 GHz. Finally it should be pointed out that capacitance C_{sub} may exhibit a very low quality factor and in order to minimize its influence on the varactor, the terminals of the substrate and the epitaxial layer are shorted together for ac signals.

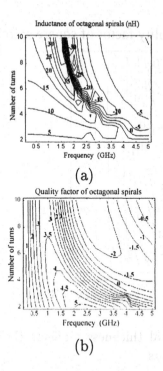

Figure 3.16: Inductance and quality factor of octagonal spiral inductors as a function of frequency and number of turns n

In a CMOS technology, both grounded and floating diodes can be fabricated as it is shown in Fig. 3.20 [17]. In such a case, a big parasitic resistance is inserted in series either with the p-type substrate or with the n-type well. In order to minimize this parasitic resistance, the diode can be surrounded by metallic contacts (strapping method).

Another technique for implementing variable capacitors that has been recently presented in literature [18] is referred to the usage of micromachining. Fig. 3.21 presents the cross-section of such a variable capacitor.

As it is depicted in the figure, the upper plate of the capacitor is a very thin aluminum plate which is kept in space and above another plate by using four mechanical springs. The air gap is approximately $1.5\mu m$ thick. When a dc voltage is applied to the capacitor, it causes an electrostatic force which pushes the upper plate towards the lower plate reducing the air gap between them and as a result, increasing the capacitance. The upper plate can be lowered by $1/3$ of its original distance from the lower plate with a 50capacitance. The use of aluminum for the fabrication of the capacitor's plates, reduces the parasitic resistances of the element and increases the quality factor.

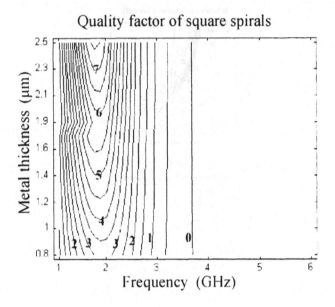

Figure 3.17: How the metal thickness t affects the quality factor of square spiral inductors

Such a structure has been experimentally tested in a VCO circuit in which tuning of 16% of the capacitance value was achieved, with a nominal capacitance value of 2pF and quality factor higher than 60 around 1 GHz.

3.4 INTEGRATED RESISTORS AND CAPACITORS FOR HIGH FREQUENCY OPERATION

3.4.1 Ohmic resistances

In analog integrated silicon technologies, ohmic resistors could be fabricated by various materials depending on the technology (bipolar, CMOS or BiCMOS). For example, resistors are made of materials that are highly doped (n+ implant or p+ implant) and exhibit very high sheet resistance (hundreds $\Omega/\square$). Finally, there are resistors that can be fabricated using the existing polysilicon layers (usually one) which exhibit relatively low sheet resistance.

Depending on the available materials and the desirable resistance value, the designer can select the best solution for each case. In high frequency circuit operation, the parasitic elements of an integrated resistors play a very important role and should be calculated and taken into account during the

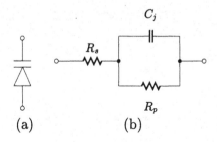

Figure 3.18: (a) Circuit symbol of varactor diode (b) Equivalent circuit model

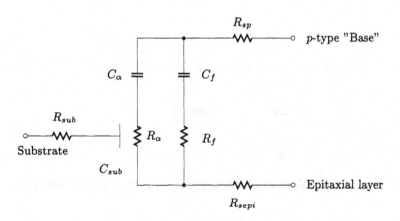

Figure 3.19: Equivalent circuit of s bipolar technology varactor

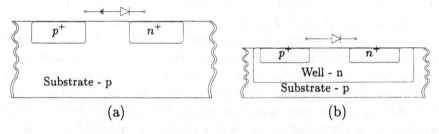

Figure 3.20: (a) Grounded diode in a CMOS technology (b) Floating diode in
a CMOS technology

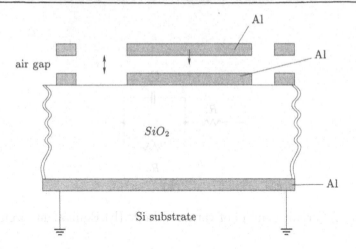

Figure 3.21: Micromachined variable capacitor

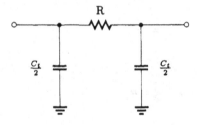

Figure 3.22: Circuit equivalent of ohmic resistor with its parasitic elements

design process. These parasitic elements consist of the parasitic capacitances to the substrate, formed by the material that the resistor is made of. There are two kinds of these capacitances: the junction surface capacitance (plate) and the fringe capacitances. For the accurate calculation of these parasitics, one has to design the resistor's layout in order to be able to calculate correctly the area and the perimeter of the device. From the geometrical data and the technology parameters, the total C_t capacitance may be calculated. This capacitance is distributed over the whole length of the ohmic resistor but up to first order approximation, it can be equally divided in two parts, each one of which can be connected to one of the resistor terminals (Fig. 3.22). In this way, the parasitic phenomena of the integrated ohmic resistors can be calculated during the circuit layout design.

3.4.2   Capacitors

In a similar way, integrated capacitors may be fabricated using various materials in an analog silicon technology depending on the available materials in each technology. For example, two polysilicon layers are available, poly/poly capacitors can be fabricated. If a technology consists of materials of high doping, poly/n+ capacitors can be fabricated as an example. In general, when circuits that operate at high frequency are designed, extreme attention should be paid to the parasitic capacitance to the substrate of every integrated capacitor. Especially if the capacitor is going to be used as a floating one (e.g. in an ac coupling), the parasitic capacitance could lead to disastrous results for the circuit operation. The thickness of the dielectric material that exists between the two plates of the capacitor and between the bottom plate of the capacitor and the substrate, virtually defines what percentage of the nominal value of the capacitor is its parasitic capacitance to the substrate.

Another parasitic element that exists in the design of integrated capacitors is the ohmic resistance of the capacitor plates which is usually significant. One acceptable solution for the reduction of these ohmic losses is the creation of multiple contacts (straps) to the terminals of the capacitor on the perimeter of its plates.

Recently, there has been a great effort towards the wide usage of low-cost CMOS technologies for the design of RF integrated circuits. The basic disadvantage of these technologies is the they rarely have the necessary layout masks for the fabrication of linear capacitors. In such a case, there are two alternatives: The creation of MOS capacitors [19] and creation of metal/insulator/metal capacitors (MIM) [20]. A digital CMOS process usually has many metalization layers (≥ 3) and the designer can benefit from the parasitic capacitance between them to create capacitors. Unfortunately, in this case, the parasitic capacitances to the substrate are a significant portion of the nominal capacitance value. On the other hand, the fact that the plates of the capacitor are made of metal, reduces the ohmic resistance resulting in a higher quality factor value.

Another loss mechanism in high frequency operation is that of the dielectric material losses which can be modeled by an ohmic resistance. The equivalent circuit model of the integrated capacitor is shown in Fig. 3.23.

R_s is the parasitic ohmic resistance of the contacts, which can vary as a function of frequency due to the skin-effect. C_{par1} and C_{par2} are the parasitic capacitances of the two plates to the substrate and R_d is the loss resistance of the dielectric. Finally, C is the nominal capacitance value of the element.

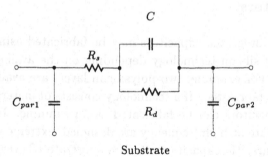

Figure 3.23: Equivalent circuit model of integrated capacitor with its parasitic elements

3.5 ELECTRICAL EQUIVALENT OF THE PACKAGE OF AN INTEGRATED CIRCUIT

The inclusion of an equivalent electrical circuit of the package in the RF IC design process has become absolutely necessary in order to achieve successful and accurate simulation results of the system performance. As a result, the selection of the appropriate package in an earlier design stage, would significantly help. Furthermore, the connection of the circuit terminals to the corresponding package terminals becomes very important because on the one hand it allows the accurate calculation of the parasitic elements (that depend on the length of the lead) and the other hand it allows the accurate prediction of the mutual inductive coupling between them.

The package elements that insert parasitic effects are mainly the leads and the bonding wires connecting the package pins with pads of the integrated circuit on the silicon die, on which the bonding wires are soldered. Fig. 3.24 depicts the structure of the package of an integrated circuit.

Leads, that can have different length according to their relative position in the package, reach near the silicon die. From that point leads are connected to the bonding wires. The other end of the bonding wires is soldered on the pads which rest on the silicon die. The pads are the on-chip terminals of the integrated circuit. Obviously, the total length of the lead defines its parasitic inductance and resistance. This also true for the bonding wires. Typical values of the above elements are presented in Table 3.1.

Fig. 3.25 presents the equivalent circuit model of two neighboring wires that are connected with the corresponding on-chip pads. Beginning from the left, there is a parasitic capacitance of the pad to the substrate and an induc-

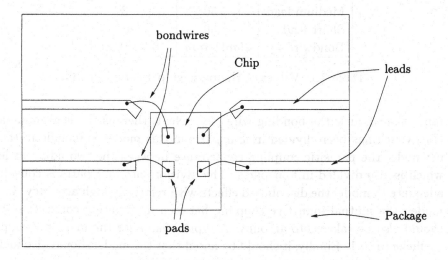

Figure 3.24: Package of an integrated circuit

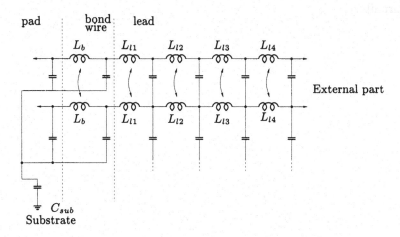

Figure 3.25: Equivalent circuit model of two neighboring terminals

	Resistance ($m\Omega$)	Inductance (nH)
Long lead	< 50	4
Medium lead	< 50	3
Short lead	< 50	< 2
Bondwire	$25 m\Omega/mm$	0.8 nH/mm

Table 3.1: Values of the package parasitic elements

tance that models the bonding wire (L_b). Next, there is the lead inductance (L_{li}) which has been divided in four parts in this model to include in the inner nodes the parasitic coupling capacitance between the two adjacent leads which is also divided in four parts. The division into four parts is considered adequate to model the distributed effects with relatively high accuracy. In this model, the mutual inductive coupling between neighboring connection wires should also be taken into account. A typical value for the magnetic coupling coefficient is 0.1. Finally, it should be noted that the model should also include the parasitic capacitance to the system ground (C_{sub}).

Accurate package models can be provided either by the fabrication industries or can be extracted by experimental measurements of structures using probe station and network analyzer (S parameters)

This concludes the presentation of the performance of active and passive integrated elements operating at high frequencies. In the following chapters, fundamental concepts and variables will be presented along with basic RF circuits.

Bibliography

[1] Q. Huang, "A MOSFET-Only Continuous-Time Bandpass Filter," *IEEE J. of Solid-State Circuits*, vol. 32, no. 2, pp. 147–158, Feb. 1997.

[2] S. Bantas and Y. Papananos, "A W-power Continuous-Time Current-Mode Filter in a Digital CMOS Process," in *Proc. IEEE 5th Int. Conference on VLSI and CAD*, Seoul - Korea, 1997, pp. 346–348.

[3] S. Pipilos, Y. Tsividis, J. Fenk, and Y. Papananos, "A Si 1.8 GHz RLC Filter with Tunable Center Frequency and Quality Factor," *IEEE J. of Solid-State Circuits*, vol. 31, no. 10, pp. 1517–1525, Oct. 1996.

[4] J.R. Long and M.A. Copeland, "The Modeling, Characterization, and Design of Monolithic Inductors for Silicon RF ICs," *IEEE J. of Solid-State Circuits*, vol. 32, no. 3, pp. 357–369, Mar. 1997.

[5] Y. Koutsoyannopoulos, Y. Papananos, C. Alemanni, and S. Bantas, "A Generic CAD Model for Arbitrarily Shaped and Multi-Layer Integrated Inductors on Silicon Substrates," in *Proc. ESSCIRC 97*, Southampton UK, Sep. 1997, pp. 320–323.

[6] Y. Koutsoyannopoulos and Y. Papananos, "A CAD Tool for Simulating the Performance of Polygonal and Multi-Layer Integrated Inductors on Silicon Substrates," in *iv Proc. IEEE 5th Int. Conference on VLSI and CAD*, Seoul - Korea, 1997, pp. 244–246.

[7] A.M Niknejad and R.G. Meyer, "Analysis and Optimization of Monolithic Inductors and Transformers for RF ICs," *IEEE CICC 1997*, pp. 375–378.

[8] H.M. Greenhouse, "Design of Planar Rectangular Microelectronic Inductors," *IEEE Trans. on Parts, Hybrids and Packaging*, vol. PHP-10, no. 2, pp. 101–109, June 1974.

[9] F.W. Grover, *Inductance Calculations*, Van Nostrand Princeton N.J. 1946, Dover Publications, 1962.

[10] R. Garg and I.L. Bahl, "Characteristics of Coupled Mictostriplines," *IEEE Trans. on Microivavc Theory and Technics*, vol. MTT-27, pp. 700–705, July 1979.

[11] H. Hasagawa, M. Furukawa, and H. Yanai, "Properties of Microstrip Line on $Si - SiO_2$ System," *IEEE Trans. on MTT*, vol. 19, pp. 869–881, Nov. 1971.

[12] H.A. Wheeler, "Transmission-Line Properties of a Strip on a Dielectric Sheet on a Plane," *IEEE Trans. on MTT*, vol. 25, pp. 631–647, Aug. 1977.

[13] E. Pettenpaul et al., "CAD Models of Lumped Elements on GaAs up to 18 GHz," *IEEE Trans. on MTT*, vol. 36, pp. 294–304, Feb. 1988.

[14] J.N. Burghartz, D.C. Edalstein, K.A. Jenkins, and Y.H. Kwark, "Spiral Inductors and Transmission Lines in Silicon Technology Using Copper-Damascene Interconnects and Low-Loss Substrates," *IEEE Trans. on MTT*, vol. 45, no. 10, pp. 1961–1968, Oct. 1997.

[15] S.M. Sze, *Physics of Semiconductor Devices*, John Wiley & Sons, 1981.

[16] M. Soyuer and R.G. Meyer, "High-Frequency Phase-Locked Loops in Monolithic Bipdar Technology," *IEEE J. of Solid-State Circuits*, vol. SC-24, pp. 787–795, June 1989.

[17] B. Razavi, "Challenges in the Design of Frequency Synthesizers for Wireless Applications," *IEEE 1997 CICC*, pp. 395–402.

[18] D.J. Young and B.E. Boser, "A Micromachine-Based RF Low-Noise Voltage-Controlled Oscillator," *IEEE 1997 CICC*, pp. 431–434.

[19] J.L. McGreary, "Matching Properties and Voltage and Temperature Dependence of MOS Capacitors," *IEEE J. of Solid-State Circuits*, vol. SC-16, no. 6, pp. 608–616, Dec. 1981.

[20] J.N. Burghartz, M. Soyuer, and K.A. Jenkins, "Microwave Inductors and Capacitors in Standard Multilevel Silicon Technology," *IEEE Trans. MTT*, vol. 44, no. 1, pp. 100–104, Jan. 1996.

CHAPTER
4

BASIC DEFINITIONS AND TERMINOLOGY

4.1 INTRODUCTION

A typical IC designer is familiar with the procedures, terminology and circuit analysis tools that are typical to a SPICE-like simulator environment. On the other hand, the microwave telecom system designers are using different methodology and tools (i.e. Smith chart), that help them proceed with their designs in an effective and proper way. The RF IC designer (especially if he comes from the low-frequency integrated circuit design field), must be familiar with the basic terminology and definitions employed in high-frequency tele-com system design in order to be able to proceed with his work. For this purpose, in the present chapter, the basic definitions and terminology employed in telecommunications systems design will be briefly reviewed. Please note that the definition of phase noise will be postponed until Chapter 7 where the oscillator circuits are presented.

4.2   DEFINITIONS

4.2.1   Input / Output Matching

For maximum power transfer and elimination of the standing waves in the various parts of a telecommunications transceiver, it is demanded that their input/output impedances exhibit equal values in a frequency range as broad as possible. The most common value for the impedance level is 50Ω. In silicon process based RF IC design, the value of the input and/or output impedance is usually dominated by an active component (bipolar or MOS transistor). The parasitic capacitances of these devices determine the equivalent impedance. It is obvious that both magnitude and phase of this impedance are frequency dependent. Impedance matching is usually requested for a particular frequency range which is the frequency range of the specific application. Therefore, input/output matching is examined only within this specific frequency range. A common measure of impedance matching is the voltage standing wave ratio (VSWR) which is defined as follows:

$$VSWR = \frac{1 + |RL|}{1 - |RL|} \tag{4.1}$$

where RL is the return loss:

$$(RL)_{dB} = 20\log\left|\frac{Z - 50}{Z + 50}\right| \tag{4.2}$$

Z is the impedance of interest. In this case, matching is performed at 50Ω. It is obvious that in the case of perfect matching, $VSWR = 1$. The procedure for evaluating $VSWR$ in SPICE-like simulators is as follows: The impedance of interest as a function of frequency is evaluated using ac analysis. $Z = Z(f) \equiv \frac{v_x}{i_x}$, as shown in Fig. 4.1. Next, using Eqs. (4.2) and (4.1), the $VSWR$ as a function of frequency is derived.

An alternative method for the evaluation of input/output impedances in a two-port network employs the S-parameters of the network as will be shown later on.

4.2.2   Conversion Gain

The power gain is the most commonly used measure in a telecommunications system. The term *conversion gain* is used to cover cases like the mixer cells where the frequency of the input signal (RF) differs from the frequency of the output signal (IF). In Fig. 4.2, the generic configuration of a two-port

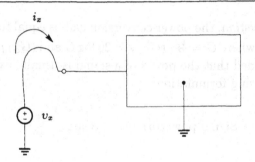

Figure 4.1: Port impedance evaluation using SPICE-like simulators

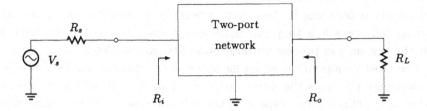

Figure 4.2: Two-port network for the definition of the power conversion gain

network is shown: At the input of the network, a signal source V_s exhibiting a source resistance R_s is connected while the output is loaded with a R_L ohmic load. The input impedance of the two-port network is R_i and its output impedance is R_o. The power conversion gain of the above configuration is defined as follows:

$$G_\alpha \equiv \frac{P_{omax}}{P_{imax}} \tag{4.3}$$

Maximum power is delivered when both input and output matching is achieved. Thus:

$$G_\alpha = \left.\frac{P_o}{P_i}\right|_{\substack{R_i=R_s \\ R_o=R_L}} \tag{4.4}$$

therefore:

$$G_\alpha = \left(\frac{V_{orms}}{V_{irms}}\right)^2 \frac{R_S}{R_L} \tag{4.5}$$

in case of both input and output matching.

It is repeated here that in telecommunications systems, matching is usually performed at 50Ω. Hence, if perfect matching is achieved both at the input and the output of the system, the power conversion gain is equal to the voltage gain G expressed in dB where $G = \frac{V_o}{V_i}$, $(G)_{dB} = 20 \log G$ and $(G_a)_{dB} = 10 \log \left(\frac{V_o}{V_i}\right)^2$.

It is also reminded that the power of a signal is usually expressed in dBm, based on the following formulation:

$$\text{Signal power}(dBm) = 10 \log \frac{\frac{V_{rms}^2}{50\Omega}}{1mW} \qquad (4.6)$$

4.2.3   Noise

The electric noise generated in an integrated circuit is a very important factor that clearly determines its behavior, especially in telecommunications applications. The reason is that this noise sets a lower bound to the signal level that the system can process without encountering problems.

The most commonly used measure for the definition of noise behavior of an electronic circuit is the *noise factor F*. The only drawback of this factor is that it is defined with respect to the internal resistor of the signal source; this resistance however is always present in the setup of a telecommunications system. The definition of F is as follows [1]:

$$F = \frac{\text{Signal-to-noise ratio at the input } \left(\frac{S_i}{N_i}\right)}{\text{Signal-to-noise-ratio at the output } \left(\frac{S_o}{N_o}\right)} \qquad (4.7)$$

The noise factor is usually expressed in dB and it is then defined as *noise figure* - NF.

The noise factor can be also defined as the ratio of the total noise at the output of the circuit to the part of the output noise contributed by the signal source resistance R_s. For example, in the case of an ideal noiseless amplifier with gain G, we have: $S_o = GS_i$ and $N_o = GN_i$ provided that the signal source resistor is the only noise source in the circuit. Thus:

$$F = \frac{S_i}{N_i}\frac{N_o}{S_o} = \frac{S_i}{N_i}\frac{GN_i}{GS_i} = 1 \qquad \text{or} \qquad 0\,\text{dB NF}$$

Therefore, the alternative definition of the noise factor becomes:

$$F = \frac{N_o}{GN_i} \qquad (4.8)$$

The electronic noise in a circuit is usually dependent on its frequency of operation and, therefore, the noise factor is a function of frequency. In many cases, such as the tuned amplifiers, the noise performance of the circuit is focused only in a narrow band Δf which is much smaller than the frequency of operation f. In this case, a different noise measure is defined, namely, the spot noise figure.

A telecommunications receiver consists of different subsystems connected in cascade. In order to evaluate the total noise factor of the receiver, the noise factor of each and every subsystem must be known. The total noise factor is then given by [2]:

$$F_{tot} = F_1 + \frac{F_2 - 1}{G_1} + \frac{F_3 - 1}{G_1 G_2} + \frac{F_4 - 1}{G_1 G_2 G_3} + \cdots \qquad (4.9)$$

where F_i are the noise factors of the various blocks and G_i are the corresponding power gain coefficients. It is obvious from Eq. (4.9), that the noise factor of the first block next to the antenna, dominates the noise performance of the receiver. This very first block is usually the Low Noise Amplifier (LNA) and must exhibit as low noise as possible and at the same time, as high gain as possible (without introducing distortion).

The evaluation of the noise figure in a circuit using SPICE, can be performed using the appropriate noise analysis as long as the circuit is linear. The procedure in this case is straightforward: SPICE calculates the total noise at the output of the circuit as a function of frequency (spot noise) as well as the gain of the circuit. The thermal noise of the source resistance is referred to the output of the circuit (using the network's transfer function) and the noise figure is evaluated. However, if the network under consideration is strongly nonlinear (e.g. VCO, mixer), the ac noise analysis in SPICE-like simulators cannot be employed anymore. In this case, different simulation tools, like transient noise analysis, must be used.

A different measure of the noise performance of a circuit is the noise temperature. The noise temperature T_n is the corresponding temperature of the source resistor R_s such that the noise contribution of this resistance calculated at the output of the circuit becomes equal to the electronic noise of the circuit. The noise measurements of the circuit are performed at temperature T. The noise factor F is related to the noise temperature T_n as follows:

$$\frac{T_n}{T} = F - 1 \qquad (4.10)$$

4.2.4 Distortion

The nonlinear nature of the electronics systems distorts the signals that are processed by them. Applying a pure sinusoidal signal at the input of such a

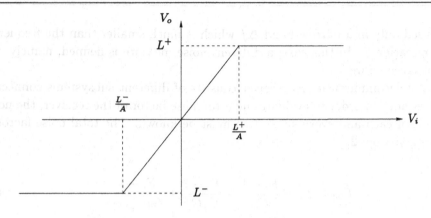

Figure 4.3: Ideal amplifier

nonlinear system results in the formation of its harmonic coefficients at the output. In telecommunications systems however, it is possible for signals from two adjacent channels to appear at the input. This causes the formation of intermodulation (IM) products at the output of the system.

a) Harmonic Distortion

The output signal V_o of a memoryless non linear system is related to its input signal v_i as follows:

$$V_o = \underbrace{V_O}_{\text{dc coefficient}} + a_1v_i + a_2v_i{}^2 + a_3v_i{}^3 + \cdots \tag{4.11}$$

Other non linear functions between input and output are also encountered as for example in the case of the ideal limiter. From Equation (4.11), two distinct cases of non linear systems are derived: the systems exhibiting strong non linearities and the systems with weak non linearities. The a_i factors of Eq. (4.11), define the nature of the system in each case.

In Fig. 4.3, the dc transfer characteristic of an ideal amplifier is shown. In this case, the amplifier exhibits a linear gain A for the input signal range $\left[\frac{L^-}{A}, \frac{L^+}{A}\right]$ and zero gain out of this region; the amplifier behaves like an ideal limiter. Such an ideal behavior, can be approximated by a series of transfer functions - some of them having weak and others having strong non linearities. Examples of such functions are shown in Table 4.1.

In the case of weakly non linear systems, the transfer function can be approximated by a linear equation for small input signals:

Type of non linearity	Analytical expression
Weak square	$v_o = v_i - 0.01v_i^2$
Strong square	$v_o = v_i - v_i^2$
Weak cubic	$v_o = v_i - 0.01v_i^3$
Strong cubic	$v_o = v_i - v_i^3$
anh	$v_o = \tanh(v_i)$

Table 4.1: Non linear transfer functions

$$v_o = a_1 \cdot v_i \tag{4.12}$$

For large signals however, the non linear nature of the system cannot be ignored anymore. Thus, if $v_i(t) = V_{iA} \cos(\omega t)$, we obtain [2,3]:

$$v_o(t) = a_1 V_{iA} \cos(\omega t) + a_2 V_{iA}^2 \cos^2(\omega t) + a_3 V_{iA}^3 \cos^3(\omega t) + \cdots \tag{4.13}$$

Performing the algebra using trigonometric properties

$$v_o(t) = \left(\frac{a_2 V_{iA}^2}{2} + \frac{3}{8}a_4 V_{iA}^4 + \cdots\right) + \left(a_1 V_{iA} + \frac{3}{4}a_3 V_{iA}^3 + \right)\cos(\omega t) + \left(\frac{a_2}{2}V_{iA}^2 + \frac{a_4}{2}V_{iA}^4 + \cdots\right)\cos(2\omega t) + \left(\frac{a_3}{4}V_{iA}^3 + \cdots\right)\cos(3\omega t) + \cdots \tag{4.14}$$

From Eq. (4.14) is derived that if at the input of a non linear system a sinusoidal waveform is applied, at the output of the system, apart from the fundamental frequency, its harmonic products appear. The amplitude of the harmonics is defined by the nature of the system's nonlinearities. Apart from the harmonic components, a *dc* component also appears at the output and causes a shifting to the quiescent point of the system. Therefore, the output exhibits harmonic distortion (HD) which is quantitatively defined by the transfer function. The harmonic distortion components HD_i are specified as follows:

$$HD_i = \frac{|b_i|}{|b_1|} \tag{4.15}$$

where b_1 is the output signal's fundamental frequency component factor and b_i is the factor of the $i - th$ harmonic component. A measure of the harmonic distortion of a non linear system is the total harmonic distortion (THD) which is defined as follows:

$$THD = \frac{\sqrt{b_2^2 + b_3^2 + \cdots}}{b_1} \tag{4.16}$$

The total harmonic distortion is expressed either in % or in dB.

b) Non harmonic distortion

Signals at various frequencies can appear at the input of a telecommunications system. Whenever more than one sinusoidal signals are simultaneously present at the input, various components - apart from the harmonics - appear at the output. Assume for instance that at the input of the system is applied a signal that is formed by two sinusoidal components: $v_i = V_{1A} \cos(\omega_1 t) + V_{2A} \cos(\omega_2 t)$. Based on the transfer function given by Eq. (4.11) and ignoring the dc components, we obtain:

$$
\begin{aligned}
v_o =\, & a_1[V_{1A} \cos(\omega_1 t) + V_{2A} \cos(\omega_2 t)] + \\
& a_2[V_{1A} \cos(\omega_1 t) + V_{2A} \cos(\omega_2 t)]^2 + \\
& a_3[V_{1A} \cos(\omega_1 t) + V_{2A} \cos(\omega_2 t)]^3 + \cdots
\end{aligned}
\tag{4.17}
$$

Consider from Eq. (4.17) only the second-order term:

$$
\begin{aligned}
a_2 v_i^2 = \, & \frac{1}{2} a_2 [V_{1A}^2 + V_{2A}^2] + \frac{1}{2} a_2 V_{1A}^2 \cos(2\omega_1 t) + \frac{1}{2} a_2 V_{2A}^2 \cos(2\omega_2 t) + \\
& a_2 V_{1A} V_{2A} \cos(\omega_1 + \omega_2)t + a_2 V_{1A} V_{2A} \cos(\omega_1 - \omega_2)t
\end{aligned}
\tag{4.18}
$$

From Eq. (4.18), it is shown that if only the second-order term of the transfer function is considered, at the output of the system appear: (i) a dc term, (ii) the second-order harmonics of the two input signals and (iii) two more components containing the sum and difference of the two input frequencies respectively. The last two terms are called second-order intermodulation products.

Following the same procedure, if from Eq. (4.17) we consider only the third-order term, the following products appear at the output:

$$
\begin{aligned}
a_3 v_i^3 = \, & a_3 V_{1A}^3 \cos^3(\omega_1 t) + 3a_3 V_{1A} V_{2A}^2 \cos(\omega_1 t)\cos^2(\omega_2 t) + \\
& 3a_3 V_{1A}^2 V_{2A} \cos^2(\omega_1 t) \cos(\omega_2 t) + a_3 V_{2A}^3 \cos^3(\omega_2 t)
\end{aligned}
\tag{4.19}
$$

From Eq. (4.19), it is shown that if only the third-order term of the transfer function is considered, at the output of the system appear: (i)the two fundamental frequencies corresponding to the two input signals, (ii) the third-order harmonics of the two input signals and (iii) all third-order intermodulation products that are formed by the summation and subtraction of one of the two fundamentals with the second harmonic component of the other ($\omega_1 \pm 2\omega_2$ and $\omega_2 \pm 2\omega_1$).

The intermodulation distortion coefficients IM_i are defined as the ratio of the amplitude of each intermodulation product to the amplitude of the fundamental component at the output of the system. The most important intermodulation distortion coefficients are the IM_2 and IM_3. If at the input of the non linear system simultaneously appear three signals, a lot of components ("beats") are present at the output. In this case, the composite output signal is called "triple beat" and the corresponding distortion coefficient TB is defined.

If one of the two signals that appear at the input of the system is modulated, a new modulated coefficient of the *unmodulated* signal is produced at the output. This type of distortion is called cross modulation distortion - CM.

Having defined the basic distortion measures in a non linear system, we can now proceed to the specification of the various criteria that determine the dynamic range of the system. The usual definition used to specify the dynamic range of a system is the ratio of the maximum distortion-free signal power to the minimum signal power that exhibits an acceptable signal-to-noise ratio. The above definition is adequate enough in the case of linear systems but usually fails to describe the phenomena that take place at the output of a non linear system where other than the fundamental frequency products also appear. Therefore, in a telecommunications system, various linearity measures can be defined in order to cover all possible sources of distortion.

4.2.5   Distortion Criteria

a) $1dB$ Compression Point - CP1

Assuming that a single tone is applied at the input of the system, the output follows the input signal according to the linear gain factor of the transfer function up to a certain point. From a particular signal level and up, the system exhibits less than the nominal gain. The output signal level where the fundamental frequency component deviates from its ideally expected value by 1 dB, is called $1dB$ *compression point*. The evaluation is usually performed on the system load and the measured quantity is signal power expressed in dBm. In Fig. 4.4, the transfer function of a non linear system is shown.

As can be seen in the above figure, the output signal is compressed beyond a particular signal level. If the linear characteristic of the system is prolonged until a certain point where the actual and the ideal characteristics differ by $1dB$, the $1dB$ compression point (CP1) is defined. CP1 is defined with respect to either the input or the output of the system in cases where the system gain is other than $0dB$.

b) Third Order Intercept - P_{3OI}

Another measure describing weak non linearities in telecommunications

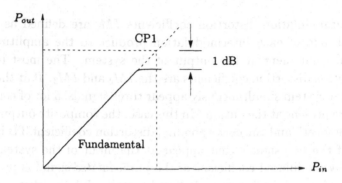

Figure 4.4: CP1 definition

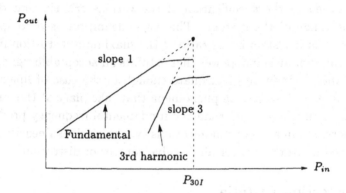

Figure 4.5: P_{3OI} definition

systems is the third order intercept P_{3OI}. The definition of P_{3OI} is graphically shown in Fig. 4.5. The power transfer characteristics of both the fundamental and the third harmonic components are plotted on the same graph.

In general, the two curves exhibit compression at different signal levels. However, if the linear segments of the two corresponding curves are prolonged, they intercept at a certain point which when projected on the horizontal axis (input power axis), define the third-order intercept. It is reminded here that at the input of the system only one signal is applied.

The above definition is valid only in simple systems [4]. For example, it assumes that the third harmonic is increased three times faster than the fundamental; this is a valid assumption for weak nonlinearities (e.g. weak cubic function). For this reason, the evaluation of the various nonlinearity measures must be carefully performed: At the system input, a small-signal excitation is applied and the corresponding output of both the fundamental and the third harmonic components is measured. This procedure is repeated

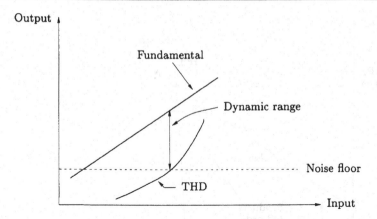

Figure 4.6: Distortion-free dynamic range definition

for slightly different input signal strengths so that the fundamental and third harmonic characteristics can be plotted. Then, the slope of these curves is evaluated and must be found approximately equal to 1 and 3 respectively so that the definition of P_{3OI} is meaningful. Following the same procedure, higher order intercepts can be also defined.

c) Dynamic Range - DR

As mentioned earlier, the term "dynamic range", refers to a quantitative measure of the behavior of a linear system. The definition of the dynamic range can be extended to non linear systems also. In this case, the definition can be altered in order to be able to describe various phenomena that take place in the operation of the system. Thus, we can concentrate on the most critical behavior that is of interest in each particular case. For example, if noise performance is of primary concern, then the dynamic range can be defined as the ratio of the maximum output signal to the output noise floor. The maximum output signal can be now defined using different criteria: For example, it could be defined as the output signal level where the harmonic distortion does not exceed a predefined value (e.g. $THD < 1\%$), or alternatively, it could be CP1.

If distortion is of primary concern, then the definition of the maximum output voltage signal becomes more stringent. In this case, the distortion-free dynamic range is defined. In Fig. 4.6, the above mentioned definition is shown: the THD is not allowed to exceed the noise floor of the system.

The definition of distortion-free dynamic range can be extended to include non harmonic distortion also: In this case, no single component at the output of the system is allowed to exceed the noise floor (spurious-free dynamic range).

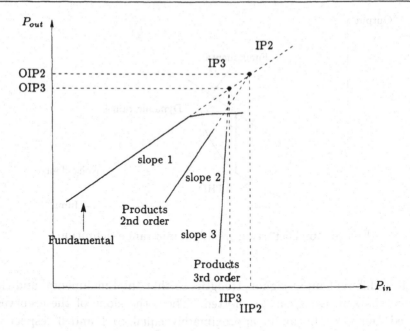

Figure 4.7: IPn definition

d) Intermodulation distortion

If a two-tones signal at frequencies ω_1 and ω_2 is applied at the input of a non linear system, intermodulation distortion products appear at the output. It was reported in the previous section that the most important intermodulation distortion coefficients are the second (IM_2) and third (IM_3) order terms.

In Fig. 4.7, the power transfer characteristic of a non linear system is shown. At the input of the system, a two-tones test signal is applied and the corresponding transfer function with respect to the output of one of the two components is plotted. Apart from the fundamental, the transfer functions of one of the second and one of the third order intermodulation products are shown.

The second order products characteristic must exhibit a slope of approximately 2. The intercept point of this curve with the linear curve of the fundamental characteristic is called $IP2$. Similarly, the third order products characteristic has a slope of approximately 3 and the corresponding intercept point with the fundamental characteristic is called $IP3$. The intercept points can refer at both the input or the output of the system and are called $IIPn$ (Input- referred Intercept Point - order n) and $OIPn$ (Output- referred Intercept Point - order n) respectively. The most commonly used measure is the $IIP3$. Please note that the $IIP3$ is different than the P_{3OI}: The former is defined by intermodulation distortion products while the latter is defined by harmonic distortion products.

Simulations and laboratory measurements of the above mentioned quantities must be carried out with extreme care: Especially when intermodulation distortion products are involved, care must be taken in the proper selection of the two tones that will be applied at the input in order that the corresponding products at the output of the system do not interfere with other unwanted signals and thus the measurement is misquoted. For this purpose, the two tones have to be close to each other. On the other hand, if the simulation is executed in the time domain using a SPICE-like simulator, the simulation time might be unacceptably long since the time step has to be very small while the duration of the simulation time is retained high.

Another issue of importance is the slope of the characteristics in various cases: The slope of the fundamental characteristic must be close to 1 while the slope of the third order harmonic or intermodulation products must be 3. The latter is closely related to the nature of the system's non linearities. Thus, in order for the designer to maintain a clear view of the various characteristics, these must be composed after executing simulations at different signal power levels starting from very low values. For this purpose, simulation tools other than the time-domain and the following FFT analysis have been developed in order to perform fast and accurate linearity evaluations. Such mathematical tools are the Volterra and power series which are briefly presented later on.

4.2.6 Definitions related to mixer cells

The mixer cells are fundamental cells in a telecommunications transceiver chain. The non linear nature of their operation and specifically the fact that the output signal is at a frequency different than that of the input signal, calls for a set of definitions useful in the design of these particular non linear cells.

a) Mixer conversion gain

As defined in Paragraph 4.2.2, the mixer conversion gain is the ratio of the output IF signal power to the available at the RF input signal power.

b) Mixer isolation

The mixer isolation is a measure of the power transferred from one port of the system to another port due to the parasitic coupling between them. The most important coupling is performed between the LO and IF gates as well as between the LO and RF gates. Consequently, the $LO - IF$ and the $LO - RF$ isolation measures are the most important ones. These measures define the fraction of the LO power that leaks to the IF and RF ports respectively. The latter is responsible for a very undesirable effect which is the emergence of a dc offset at the IF output that is caused by the mixing of an RF signal at the local oscillator frequency with the LO signal itself.

c) Mixer noise figure

Two types of noise figures are defined in mixers: the single sideband (SSB) and the double sideband (DSB) [5]. The DSB noise figure includes both the noise power of the IF component and the noise power of the image component. The SSB noise figure contains only the noise power of the IF component while the noise power of the image component is rejected. In order for this to be achieved, an image-reject filter has to be applied at the RF input of the mixer. Details on the definition of the image frequency and the image-rejection mixers will be presented in Chapter 6.

4.2.7   Volterra Series and Power Series

The analysis of non linear phenomena in weakly non linear systems is of paramount importance for the operation of telecommunication transceiver systems. The most powerful mathematical tools used in this case are the power series and the Volterra series. The analysis employing power series is straightforward but assumes a simplification that is usually not valid. More specifically, it assumes that the circuit is non linear and memoryless. The more general case of Volterra series provides a satisfactory solution to the problem. The reader is referred to [6-9] for a detailed presentation of the above mentioned approach.

Power series

The memoryless non linear transfer function can be described by the following equation:

$$v_{out}(t) = \sum_{n=1}^{N} \alpha_n v_{in}{}^n(t) \tag{4.20}$$

The weakly non linear nature of the transfer function can be described by the value of N in the series expansion. Usually N must take a value in between 3 and 5 in order for the mathematical analysis of the system to be feasible. If the input signal contains more than one signals:

$$v_s(t) = \frac{1}{2} \sum_{\substack{q=-Q \\ q\neq 0}}^{Q} V_{S,q} \cdot \exp(j\omega_q t) \tag{4.21}$$

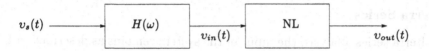

Figure 4.8: Power series model of a non linear system

then the n-th order term at the output of the system can be described based
on Eq. (4.20) as follows:

$$\alpha_n v_{out}{}^n(t) = \alpha_n \Big[\frac{1}{2} \sum_{q=-Q}^{Q} V_{S,q} H(\omega_q) \exp(j\omega_q t)\Big]^n =$$

$$\frac{\alpha_n}{2^n} \sum_{q_1=-Q}^{Q} \sum_{q_2=-Q}^{Q} \cdots \sum_{q_n=-Q}^{Q} V_{S,q_1} \cdot V_{S,q_2} \cdots V_{S,q_n} \cdot H(\omega_{q_1}) \cdot$$

$$\cdot H(\omega_{q_2}) \cdots H(\omega_{q_n}) \cdot \exp[j(\omega_{q_1} + \omega_{q_2} + \cdots + \omega_{q_n})t]$$

(4.22)

where $H(\omega_{q_i})$ is the transfer function of the linear system preceding the
model of Fig. 4.8 that describes the power series. It is assumed in the above
analysis that no dc component appears at both the input ($q \neq 0$ in Eq. (4.21))
and the output of the system.

However please note that in practice, there is always a small dc component
at the output and this causes a shifting in the quiescent point of the circuit.
If the output dc component is strong (for example in class B amplifiers), the
analysis based on power or Volterra series cannot be applied anymore.

The complete output response is formed by adding all factors of the form
of Eq. (4.22) based in Eq. (4.21). If the number of the input components
is two ($Q = 2$) and the weak non linearities are defined by the inequality
$N \leq 3$, then the power series analysis gives all the harmonic and non harmonic
components at the output. It is obvious that, depending on the values of Q
and N, a large number of components can appear at the output. It is worth
noting however that it is not easy to define the order of the product from the
frequency at which it appears. For example, a signal at frequency $2\omega_1 - \omega_2$
can be produced by mixing the components $\omega_1 + \omega_1 - \omega_2$ (order 3) and/or the
components $\omega_1 + \omega_1 + \omega_1 - \omega_1 - \omega_2$ (order 5).

Volterra Series

In Volterra series analysis, the input of the system remains as described in Eq. (4.21) and the output is given by the following equation:

$$v_{out}(t) = \sum_{n=1}^{N} \frac{1}{2^n} \sum_{q_1=-Q}^{Q} \sum_{q_2=-Q}^{Q} \cdots \sum_{q_n=-Q}^{Q} V_{S,q_1} \cdot V_{S,q_2} \cdots V_{S,q_n} \cdot$$
$$H_n(\omega_{q_1}, \omega_{q_2} \ldots \omega_{q_n}) \cdot \exp[j(\omega_{q_1} + \omega_{q_2} + \cdots + \omega_{q_n})t] \tag{4.23}$$

The only difference from the power series case is the introduction of a non linear function $H_n(\omega_{q_1}, \omega_{q_2}, ., ., \omega_{q_n})$ in the transfer function. In the general case, n linear transfer components $H(\omega_{q_i})$ (Eq.(4.22)) are involved in the definition of the transfer function using power series. Knowing the exact nature of the non linear function H_n, all the output components can be calculated - exactly as in the power series case. From Eqs. (4.22) and (4.23), the power series are a special case of the Volterra series, where

$$H_n(\omega_{q_1}, \omega_{q_2}, \ldots, \omega_{q_n}) = \alpha_n H(\omega_{q_1}) H(\omega_{q_2}) \cdots H(\omega_{q_n}) \tag{4.24}$$

The non linear transfer function can be found using a method called harmonic input method. This procedure is quite similar to the evaluation of the transfer function $H(\omega)$ of a linear system in the frequency domain. In order to find the n-th order output component, the following signal is applied to the input of the system

$$v_s(t) = \exp(j\omega_1 t) + \exp(j\omega_2 t) + \cdots + \exp(j\omega_n t) \tag{4.25}$$

(n phasors of amplitude 1) and from Eq. (4.23), the n-th order output component at the frequency $\omega_1 + \omega_2 + \cdots + \omega_n$ is evaluated:

$$v_{out}(t)|_{\omega=\omega_1+\omega_2+\cdots+\omega_n} = n! H_n(\omega_1, \omega_2, \ldots, \omega_n) \exp[j(\omega_1 + \omega_2 + \cdots + \omega_n)t] \tag{4.26}$$

The above expression for $v_{out}(t)$ is replaced in the system equations keeping only the n-th order terms. Thus, the non linear function H_n can be evaluated algebraically. For this purpose, various computer simulation programs have been developed during the past years for the evaluation of non linear circuits. These CAD tools, like Hewlett Packard's EEsof and Cadence's Spectre RF, are using the above mentioned methods for fast and accurate calculation of the output spectrum without employing the computer and time consuming analysis in the time domain and the FFT thereinafter.

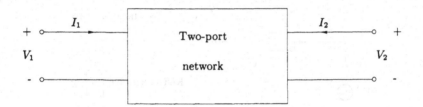

Figure 4.9: Two-port network

4.3   S PARAMETERS

The operation of linear and weakly non linear systems can be described using parameters that are evaluated at the system ports (n ports in general), without any knowledge of the contents of the system. More specifically, the S-parameters (scattering parameters) are employed in the design of microwave circuits due to the fact that they can be easily measured at high frequencies. The evaluation of parameters of other types calls for open and short circuiting of the input and output ports of the network - a task that is extremely difficult to accomplish at high frequencies.

On the other hand, the S-parameters can be measured by merely employing source and load impedances that are necessary to fulfil the matching constraints of high-frequency operation. On the contrary, a short or open circuit at the system ports can lead to oscillations.

The analysis presented in this section is limited to two-port networks. The general case of n-port networks is obvious. In Fig. 4.9, a two-port network with the voltage and current signals at its ports is drawn. The most commonly used parameters are the H, Y and Z parameters as defined below:

$$V_1 = h_{11}I_1 + h_{12}V_2 \tag{4.27}$$

$$I_2 = h_{21}I_1 + h_{22}V_2 \tag{4.28}$$

$$I_1 = y_{11}V_1 + y_{12}V_2 \tag{4.29}$$

$$I_2 = y_{21}V_1 + y_{22}V_2 \tag{4.30}$$

$$V_1 = z_{11}I_1 + z_{12}I_2 \tag{4.31}$$

$$V_2 = z_{21}I_1 + z_{22}I_2 \tag{4.32}$$

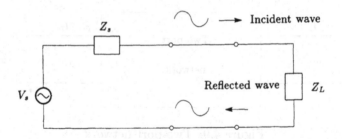

Figure 4.10: Power transfer

The only difference between the above parameters is in the selection of the independent and dependent variables in each case. In order to evaluate any of them, the corresponding port must be open or short circuited and the measurement must be performed at the other port. For example, in order to evaluate the h_{11} parameter, the output port of the network is short circuited:

$$h_{11} = \left.\frac{V_1}{I_1}\right|_{V_2=0} \tag{4.33}$$

Thus, the h_{11} parameter is simply the input impedance of the network as shown in Eq. (4.30) when the output port is short circuited.

At high frequencies however, the total voltage and current quantities are not adequate enough for the network characterization. In Fig. 4.10, a power signal source with a source impedance Z_s is shown. This source, delivers a signal to the load impedance Z_L through transmission lines. Part of the wave arriving at the load is reflected back to the source and if $Z_s \neq Z_o$ (Z_o is the characteristic impedance of the transmission line), it is reflected again to the load. In this way, standing waves are formed. Usually, the characteristic impedance of the transmission lines exhibits a value of 50Ω. The voltage at any point in the transmission line is always the sum of the incident and the reflected waves [10]:

$$V_t = E_{inc} + E_{ref} \tag{4.34}$$

$$I_t = \frac{E_{inc} - E_{ref}}{Z_o} \tag{4.35}$$

In Eq. (4.35), the corresponding definition for the total current at a certain point is given. If in Fig. 4.10, a two-port network is inserted (Fig. 4.11), travelling waves from/to both ports of the network will occur. For example,

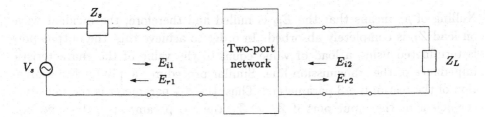

Figure 4.11: Two-port network for the definition of S parameters

E_{r2} consists of the part of E_{i2} that is reflected from the output port and of the part of E_{i1} that travels through the network. Similar relations hold for the rest of the waves. Thus:

$$V_1 = E_{i1} + E_{r1}, V_2 = E_{i2} + E_{r2} \qquad (4.36)$$

$$I_1 = \frac{E_{i1} - E_{r1}}{Z_o}, I_2 = \frac{E_{i2} - E_{r2}}{Z_o} \qquad (4.37)$$

Based on the above, the following parameters are defined:

$$a_1 = \frac{E_{i1}}{\sqrt{Z_o}}, \qquad a_2 = \frac{E_{i2}}{\sqrt{Z_o}} \qquad (4.38)$$

$$b_1 = \frac{E_{r1}}{\sqrt{Z_o}}, \qquad b_2 = \frac{E_{r2}}{\sqrt{Z_o}} \qquad (4.39)$$

The new parameters refer to travelling waves and not to total voltage and current quantities and can be related through the S parameters:

$$b_1 = S_{11}a_1 + S_{12}a_2, \qquad (4.40)$$

$$b_2 = S_{21}a_1 + S_{22}a_2, \qquad (4.41)$$

S-parameters measurements are performed in a similar to the rest of the parameters manner. For example, the S_{11} parameter measurement is performed by putting $a_2 = 0$.

$$S_{11} = \left.\frac{b_1}{a_1}\right|_{a_2=0} \qquad (4.42)$$

Nulling of a_2 means that the E_{i2} is nulled and therefore, the incident wave on load Z_L is completely absorbed. In order to achieve this, the output port is terminated using a load of value equal to the value of the characteristic impedance of the transmission line. Similar procedures apply for the evaluation of the rest of the S parameters. Thus, the S_{11} parameter is the reflection coefficient at the input port if $Z_L = Z_o$, the S_{22} parameter is the reflection coefficient at the output port if $Z_s = Z_o$, the S_{21} parameter is the forward transmission gain if $Z_L = Z_o$ and the S_{12} parameter is reverse transmission gain if $Z_s = Z_o$.

The S_{11} and S_{22} coefficients are directly related to the input and output impedances of the two-port network. For example:

$$S_{11} = \frac{Z_1 - Z_o}{Z_1 + Z_o} \tag{4.43}$$

where Z_1 is the input impedance at port 1.

The above relationship between the impedance and the reflection coefficients is the basis on which the Smith chart is formed. A very interesting relationship is the one between the S parameters and the power transfer in the system under consideration [11]:

$|S_{11}|^2$ is the ratio of the power reflected from the network input to the power delivered to the network input. $|S_{22}|^2$ is the ratio of the power reflected from the network output to the power delivered to the network input. $|S_{21}|^2$ is the ratio of the power delivered to a load Z_o to the power available from a source impedance (forward power gain). $|S_{12}|^2$ is the reverse power gain.

Bibliography

[1] P.R. Gray and R.G. Meyer, *Analysis and Design of Analog Integrated Circuits*, John Wiley, 1984.

[2] U.L. Rohde and T.T.N. Bucher, *Communications Receivers: Principles and Design*, Mc Graw Hill, 1994.

[3] Y. Tsividis *Telecommunications Electronics*, Lecture Notes - in Greek, 1993.

[4] B. Gilbert, *Mixer Fundamentals and Active Mixer Design*, RF IC Design for Wireless Communication Systems, Lausane, Switzerland, 1995.

[5] A. Dao, *Integrated LNA and Mixer Basics*, National Semiconductor Application Note 884, 1993.

[6] D.D. Weiner and J.F. Spina, *Sinusoidal Analysis and Modeling of Weakly Nonlinear Circuits*, Van Nostrand, New York, 1980.

[7] J.J. Bussgang, L. Ehrman, and J.W. Graham, "Analysis of Nonlinear Systems with Multiple Inputs," *Proc. IEEE*, vol. 62, pp. 1088, 1974.

[8] M. Schetzen, *The Volterra and Wiener Theories of Nonlinear Systems*, John Wiley, New York, 1980.

[9] V. Volterra, *Theory of Functionals and of Integral and Integro-Differential Equations*, Dover, New York, 1959.

[10] *S-Parameter Design*, Hewlett Packard, Application Note 154.

[11] *S-Parameter Techniques*, Hewlett Packard, Application Note 95-1.

CHAPTER
5

LOW NOISE AMPLIFIERS

5.1 INTRODUCTION

The Low Noise Amplifier (LNA) is one of the most important and crucial parts in a telecommunications transceiver. It is the first active circuit in the receiver part following the antenna as shown in Fig. 5.1 for the superheterodyne architecture. Due to its location in the receiver chain, it dominates the noise performance and the VSWR of the complete system. Therefore, a LNA must exhibit low noise figure, high gain, good linearity performance and input VSWR as low as possible.

The rapid evolution of the mobile communication systems during the past few years, calls for the continuous improvement in performance along with the minimization of power consumption. This combination poses a real challenge to the RF IC designer. Many successful LNA designs appear in the literature;

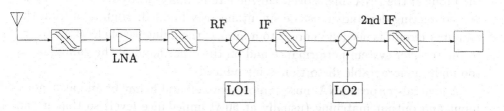

Figure 5.1: Schematic diagram of a superheterodyne receiver

some employing bipolar, some CMOS and others employing BiCMOS technologies. The application domain covers a frequency span up to 6 GHz. For example, designs have been reported for mobile and wireless communications systems up to 2 GHz (GSM, DCS, DECT), others for GPS applications at 1.6 GHz and more recently, designs in SiGe technologies operating at 5.8 GHz for wireless LAN applications. On top of everything, new applications emerge like the satellite television in the Ku band. All these markets push towards the minimization of the cost which in turn leads to the utilization of the most cost effective integrated technology, namely, the CMOS technology,

A rapid classification of the existing LNA solutions [1], reveals that for a SiGe HBT technology, the optimum performance is: NF=0.95 dB, gain around 11 dB and power consumption 2 mW. Similar performance can be found in GaAs designs. On the contrary, CMOS implementations exhibit inferior performance ($NF > 2dB$, power consumption around 10 mW). The extremely low cost however, makes the CMOS implementations a very attractive solution.

In the present chapter, it will be attempted to present the most important LNA implementations in each technology. Moreover, apart from the presentation of the broadband amplifiers, a brief overview of the tuned amplifiers will be attempted. This type of integrated circuits have recently appeared in the literature thanks to the utilization of on-chip spiral inductors.

5.2   LOW NOISE AMPLIFIERS IN BIPOLAR TECHNOLOGIES

The main target of the LNA circuits is to amplify the low power RF signals that appear at the antenna in order for the rest of the receiver subsystems to be able to process them efficiently. It is demanded that the LNA exhibits the highest possible dynamic range so that it will be able to process any signal that arrives at its input. Moreover, in order to minimize the intermodulation distortion products at the output, the LNA must also exhibit high linearity. This is especially critical if at the antenna arrives a strong RF signal. Strong signals at the input of the LNA might overload the circuit and lead to distortion. Thus, two conflicting demands must be simultaneously fulfilled: High gain from the amplifier circuit is needed in order to minimize the influence of electronic noise to the receiver system performance and on the other hand, if the gain value is too high, unacceptable distortion is introduced.

A general-purpose LNA must exhibit a broadband behavior and also, good input and output matching (usually at 50 Ω impedance level) so that it can

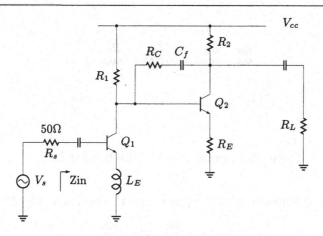

Figure 5.2: Bipolar LNA

be connected to external filters (i.e. SAW, ceramic) if necessary.

5.2.1 Two-stage Amplifier

The basic topology of a low noise amplifier implemented in a bipolar technology, is shown in Fig. 5.2 and it comprises two stages [2]. The two stages are needed to provide the necessary gain of the amplifier, especially in cases where a high-speed technology is not available. Moreover, the second stage is employed to provide good output matching while the first stage is designed to achieve both input matching and low noise operation. The noise figure reduction is performed by utilizing the bond-wire inductance at the emitter of the transistor of the first stage.

The small-signal input impedance of the bipolar transistor Q_1 becomes negligible at high frequencies due to the fact that the C_π capacitance in combination with the Miller capacitance that comes from the C_μ, annihilate the r_π resistance. Thus, at the input of the Q_1 transistor, only the base resistance r_b appears which in turn is very small (around 10 Ω) since the Q_1 transistor is a large area device. The size of Q_1 is imposed by the low noise performance specification of the amplifier. The existence of the (parasitic) L_E, leads to an increase in Z_{in} towards the 50 Ω value without however increasing the noise figure of the LNA.

If the amplifier is operated at frequencies around 1 GHz, the bipolar devices operate quite far away from their corner frequency f_b, (i.e. the frequency at which the current gain starts decreasing, mainly due to C_π). This is the cause

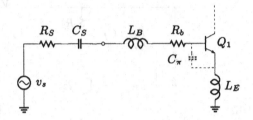

Figure 5.3: Input stage of the bipolar LNA

of the capacitive behavior of the input current. The current gain is therefore:

$$\beta = -j\frac{g_m}{\omega C_\pi} = -j\frac{\omega_T}{\omega} \tag{5.1}$$

For the circuit topology of Fig. 5.2, we have:

$$v_{L_E} = i_E j\omega L_E = i_B \beta j\omega L_E = i_B\left[-j\frac{\omega_T}{\omega}\right]j\omega L_E = i_B\omega_T L_E \tag{5.2}$$

This means that the parasitic inductor L_E behaves like an ohmic resistor of value $\omega_T L_E$ looking from the base terminal. Thus, by properly selecting the inductance value of L_E, perfect input matching can be achieved at a particular frequency.

The voltage drop on the base resistance is:

$$v_{r_b} = i_B r_b \tag{5.3}$$

By combining Eq. (5.2) and Eq. (5.3), we get:

$$v_{r_b} = i_E j\omega \frac{r_b}{\omega_T} \tag{5.4}$$

This means that the base resistance looking from the emitter terminal, behaves like a purely inductive element.

Based on the above analysis, the design of the input stage of the LNA for optimum impedance matching is performed. In Fig. 5.3, the corresponding schematic diagram is drawn.

The C_S capacitor is the coupling capacitor. The base resistance r_b is noted separately and the inductance L_B can be implemented on-chip or by an external PCB inductor. L_E is the bondwire inductance. The capacitance C_π sketched with broken lines in the figure, is the small-signal model capacitance. In order to achieve matching, the following equations must hold:

$$\omega L_B = \frac{1}{\omega C_\pi} \tag{5.5a}$$

$$R_S = r_b + \omega_T L_E \qquad (5.5b)$$

This can be achieved by properly selecting the inductors L_B and L_E. If L_B is implemented by an external inductor, then the input lead/bond wire parasitic inductance must be also taken into account. It is also noted that if $r_b = R_S$ (in which case the L_E is not needed), the noise figure of the LNA cannot exceed the value of 3 dB. However, modern communication systems demand LNA designs with NF as close to 2 dB as possible. In order then to meet this specification, the base resistance must be as small as possible. This calls for bipolar devices with large areas and, therefore, exhibiting large parasitic capacitances. These capacitances in turn define the speed of operation of the device and thus, they limit the bandwidth of the LNA. This is why the high-speed bipolar technologies are the noisiest ones. The proper selection of transistor Q_1 is one of the most critical steps towards the design of a high quality LNA of the type of Figure 5.2. An approximate formula for the evaluation of the noise figure of this topology is [2]:

$$NF = 1 + \frac{r_b}{R_S} + \frac{1}{2g_m R_S} + \frac{g_m R_S}{2|\beta|^2} + \frac{g_m R_S}{2\beta_0} \qquad (5.6)$$

The role of r_b in the noise figure of the amplifier is now evident from the above relationship.

Power dissipation is of paramount importance in the performance of an LNA. The current dissipation of the first amplifier stage defines the gain and the noise performance of the circuit while the current consumption of the second stage defines the linearity performance of the circuit. The feedback network (R_f, C_f) defines the output matching.

A major drawback of the present topology is the large gain variation due to the voltage supply variation and the bias resistors (R_1, R_2) variation. In order to overcome this problem, the collector currents of Q_1 and Q_2 must become proportional to absolute temperature - PTAT. This can be achieved by using PTAT supplies that can be internally generated. The complete schematic of the LNA (including the on-chip bias circuitry) is shown in Fig. 5.4 [2].

As shown in Fig. 5.4, at the base of Q_4 a voltage equal to $2V_{BE} + V_{PTAT}$ is applied through the unity gain amplifier and an RC filter used to reject the high-frequency noise. The dc loop forces $V_{CE1} = V_{BE2} + 2V_{BE} - V_{BE4} - V_{BE2} \approx V_{BE2}$. Thus:

$$I_{C_1} = \frac{V_{PTAT} + 2V_{BE} - V_{BE4} - V_{BE2}}{R_1} \approx \frac{V_{PTAT}}{R_1} \qquad (5.7)$$

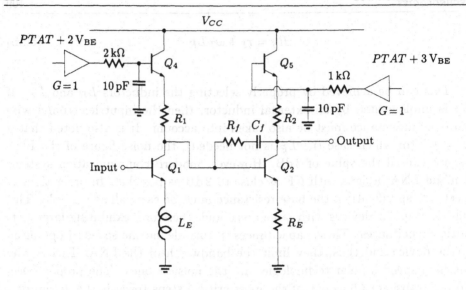

Figure 5.4: Complete LNA schematic

A similar technique is followed in the biasing of Q_2. The biasing of the LNA is stabilized through the dc feedback loop shown in Fig. 5.5. This loop provides sufficient feedback gain at low frequencies in order to achieve biasing stabilization without affecting the high frequency operation of the LNA or loading the output stage of the circuit.

If for the implementation of the above topology, a BiCMOS technology is available, the complicated dc bias PTAT circuitry can be eliminated and the bias resistors can be replaced by pMOS devices. The pMOS transistors approximately operate like constant current sources providing constant bias current to the transistors Q_1 and Q_2. Thus, in the circuit of Fig. 5.5, only the dc feedback loop remains for the proper biasing of the base terminals of the bipolar transistors.

The circuit described in this section has been successfully implemented [2] for operation in the 900 MHz band.

Following, the simulation results of the presented LNA topology are reported. The results refer to a BiCMOS implementation with an f_T value for the bipolar devices of approximately 11 GHz. pMOS transistors have been used as loads. The input transistor is a low-noise device with the lowest possible C_π. The parasitic capacitances of the integrated ohmic resistors have been also taken into account in the simulations. Moreover, it is worth noting that the package parasitics play a very important role in the operation of the LNA since this type of amplifier is extremely sensitive to them. Package parasitic

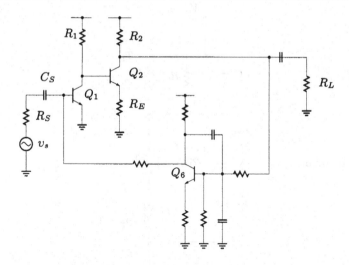

Figure 5.5: dc circuit part of the LNA

Parameter	Simulated Value
Frequency of Operation	900 MHz
Gain	12 dB
Noise Figure	2.35 dB
Input VSWR	1.2
Output VSWR	1.2
OIP3	+7 dBm
Current Consumption	8 mA
Supply Voltage	3 V

Table 5.1: Performance of the LNA of Fig. 5.2

devices must be a *part* of the circuit topology in the simulations. The LNA performance is even affected by the pin assignment. Taking all the above into account, the simulation results of the LNA are briefly reported in Table 5.1.

5.2.2 Amplifier Employing a Transformer

A very interesting low noise amplifier topology using an integrated transformer, is presented in [3]. The circuit topology is shown in Fig. 5.6. The transformer T_1 poses a negative feedback loop by inductively coupling the collector and emitter of the transistor Q_1. Thus, the linearity of the circuit is improved and gain stability against temperature and other variations is ensured.

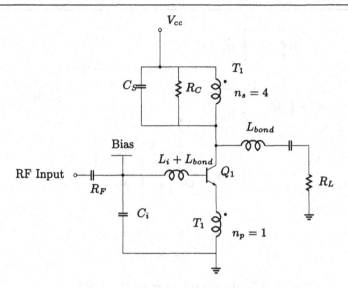

Figure 5.6: LNA using a transformer

The connection of the transformer winding in the collector and emitter terminals of the transistor, provides voltages to these terminals close to the supply voltage due to the low ohmic losses of the transformer. Thus, the proposed circuit topology, can operate with supply voltages as low as 0.9 V.

Input matching and low noise figure can be achieved for this amplifier by connecting an inductor in series with the emitter terminal of transistor Q_1 [4]. However, for this purpose, a large inductance value is demanded. This would lead to the implementation of a two-stage amplifier in order to achieve the wanted gain from the LNA. A two-stage topology inevitably leads to an increase in power consumption. An alternative solution has been applied in the circuit of Fig. 5.6 where the devices C_i and L_i are used for both input matching and low noise design purposes.

In this particular circuit, the dominant noise sources are the thermal noise of the transistor's base resistance and the shot noise due to the collector bias current. The R_C resistor that acts as the collector's load and is used for output matching purposes, does not contribute significantly to the amplifier's total noise because it is close to the amplifier's output. As in the previous case, careful selection of the transistor is demanded in order to optimize both power consumption and noise performance. In common emitter configurations, the noise figure is proportional to the following product [5]:

$$NF \propto g_m r_b \left(\frac{f}{f_T}\right)^2 \tag{5.8}$$

From Eq. (5.8), the compromise between noise and bandwidth i.e. between the value of r_b and the value of f_T is evident. If both the input and the output are matched, the amplifier's power gain is approximately given by the following expression:

$$\text{power gain} = |S_{21}|^2$$

$$= \left| \frac{-g_m R_L}{A_{BJT} + g_m R_L \left[\frac{1}{n} + j\omega r_b C_\mu(\frac{1}{n}+1) - \omega^2 L_i' C_\mu(\frac{1}{n}+1)\right]} \right|^2 \quad (5.9)$$

where $L_i' = L_i + L_{bond}$, n is the transformer's turn ratio and A_{BJT} is:

$$A_{BJT} = 1 + j\omega r_b(C_\pi + C_\mu) - \omega^2 L_i'(C_\pi + C_\mu) \quad (5.10)$$

For low frequency operation, the parasitic capacitances of the transistor can be ignored and thus, the power gain becomes:

$$|S_{21}|^2 \approx \left| \frac{-g_m R_L}{1 + g_m R_L(\frac{1}{n})} \right|^2 \quad (5.11)$$

From Eq. (5.11) it is obvious that the power gain of the LNA is proportional to the square of the turn ratio n (for high $g_m R_L$ values). The same attribute holds also for operation at high frequencies due to the double resonance (Eq. (5.9) and Eq. (5.10)): The first resonance is the series resonance at the input and the second is owed to the combination of the feedforward due to the collector-base capacitance (C_μ) with the feedback action of the transformer. At the resonance frequency of ($L_i + L_{bond}$) with ($C_\pi + C_\mu$), the term A_{BJT} becomes negligible in Eq. (5.9). In the same equation, the terms containing the capacitance C_μ are neglected and thus, the power gain in Eq. (5.9) is also depended on the square of n. In the implementation described in [3], a value of $n = 4$ was selected for operation of the LNA at 1.9 GHz and a power gain of 12 dB. In this circuit, the measured noise figure was 2.8 dB. In Table 5.2, the measurement results are summarized. The LNA was implemented in a BiCMOS 0.8 μm process with $f_T = 11$ GHz for the bipolar devices.

An alternative implementation of the above circuit, implemented in a high-speed bipolar technology ($f_T = 25$ GHz) for DECT applications, can be found in [6].

5.2.3   Broadband Two-Stage Bipolar Amplifier

Next, a broadband amplifier implemented in a bipolar technology is presented. The particular circuit, exhibits high power consumption compared to other existing topologies. The simplified schematic of the circuit is shown in Fig. 5.7.

Parameter	Measured Value
Frequency of Operation	1.9 GHz
Gain	9.5 dB
Noise Figure	2.8 dB
Input VSWR	1.2
Output VSWR	1.4
IIP3	-3 dBm
Current Consumption	2 mA
Supply Voltage	1.9V

Table 5.2: Performance of the LNA of Fig. 5.6

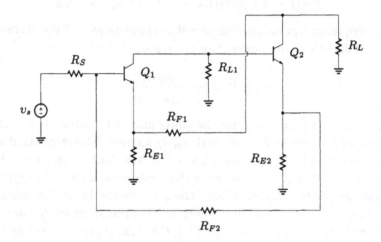

Figure 5.7: Broadband amplifier

The amplifier [7], uses multiple feedback loops [8] in order to achieve broadband operation, sufficient gain due to the existence of the two stages and, simultaneously, low noise figure and input/output matching. The amplifier's gain is dominated by the ratio $(R_{F1} + R_{E1})/R_{E1}$ which is a series-shunt feedback configuration.

The shunt-series feedback formed by R_{F2} and R_{E2}, provides matching without the need for any extra component as for example the ohmic resistor connected in parallel to the amplifier's input which would inevitably increase the noise figure of the circuit. For optimum noise performance, R_{E1} and the base resistance r_b of Q_1 must be minimized and R_{F2} maximized. It is obvious that the resistors' values define the biasing conditions of the amplifier and therefore, must be carefully selected in order to find the optimum solution that would lead to minimum power consumption while retaining the performance of the amplifier. In this case, as in the previous designs, careful selection of the input transistor is demanded; multiple-base devices are usually preferred in order to decrease the value of r_b.

The output stage of the configuration is implemented using a Darlington pair which increases the feedback gain. The proper selection of the ground terminals is also of paramount importance in order to avoid parasitic signal paths in the packaged integrated circuit that could lead to oscillations.

5.3 LOW NOISE AMPLIFIERS IN MOS TECHNOLOGIES

The implementation of MOS RF integrated circuits is now feasible thanks to the progress in IC design technology during the past years. The obvious advantage in this case is the reduced fabrication cost and the compatibility with the baseband processing circuitry that can lead to higher levels of integration. The latter leads to reduced number of components in the transceiver and therefore, both the cost and the size of the system are diminished. The MOS technology is thus a technology that can lead in the near future to one-chip transceiver solutions.

5.3.1 LNA Employing a Current Reuse Technique

A very interesting LNA implementation in a 0.5 μm CMOS technology for NADC (North American Digital Cellular) applications, can be found in [9], [10]. The basic circuit schematic is shown in Fig. 5.8. In this figure, the biasing network is simplified and is represented by the resistor R_{bias}, while input matching is performed through the inductors L_G and L_S, according to the following formulation:

$$\omega^2 C_{gs}(L_G + L_S) \approx 1 \tag{5.12a}$$

$$\text{thus} \quad L_S \approx \frac{R_S C_{gs}}{g_m} \tag{5.12b}$$

Based on the above equations, the noise figure of the amplifier is given by Eq. (5.13):

$$F \approx 1 + (8\omega^2 C_{gs}^2 R_S)/(3g_m) \tag{5.13}$$

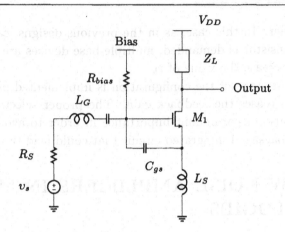

Figure 5.8: CMOS LNA

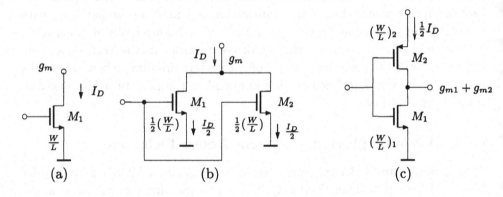

Figure 5.9: Current reuse method

The equation Eq. (5.13) shows that it is feasible to achieve a noise figure around 2 dB from this topology at the extra cost of increased drain current. However, in order to maintain low current levels for consumption purposes, the current reuse method presented in Fig. 5.9 is applied. The target is to achieve the desirable transconductance and $\omega_T = g_m/C_{gs}$ frequency values with the lowest possible current consumption.

The transistor M_1 of Fig. 5.9(a), is split in two devices M_1 and M_2 connected in parallel and having half the $\frac{W}{L}$ ratio each with respect to the transistor of Fig. 5.9(a). Thus, the current in each transistor of Fig. 5.9(b), will be $\frac{I_D}{2}$ i.e. half the original value. Next, in Fig. 5.9(c), the transistor M_2 is replaced by a pMOS device and now the transistors M_1 and M_2 are connected in series (CMOS inverter configuration). The total transconductance is now $g_{m1} + g_{m2}$ which is the sum of the transconductances of the two transistors.

Parameter	Measured Value
Frequency of Operation	900 MHz
Gain	15.6 dB
Noise Figure	2.2 dB
IIP3	-3.2 dBm
CP1	-15.2 dBm
Supply Voltage	2.7V
Power Dissipation	20 mW

Table 5.3: Performance of the LNA of Fig. 5.8

If a proper $\frac{W}{L}$ ratio is selected so that $\left(\frac{W}{L}\right)_1 = \left(\frac{W}{L}\right)_2 = \frac{1}{2}\left(\frac{W}{L}\right)$, the total capacitance C_{gs} of the topology of Fig. 5.9(c) is $C_{gs1} + C_{gs2}$. This value is approximately equal to the C_{gs} value of the transistor of Fig. 5.9(a). However, the total transconductance in the case of Fig. 5.9(c) is less than the transconductance of the transistor in Fig. 5.9(a) due to the lower carrier mobility of the pMOS devices. This reduction in transconductance value leads to a slight increase in noise figure but the current consumption has been halved thanks to the application of the current reuse technique. For example, if $\mu_p = 0.5\mu_n$, the total transconductance of the circuit of Fig. 5.9(c) is 0.85 times the transconductance of the circuit of Fig. 5.9(a). This leads to an increase of the noise figure value by 0.2 dB.

In order for the circuit of Fig. 5.8 to meet the specs of mobile communication systems at 900 MHz, it is demanded that two stages using the current reuse technique are connected in cascade. This way it is easier to meet the specifications while achieving output impedance matching in case the LNA must drive an off-chip load. The measurement results for this structure are summarized in Table 5.3.

5.3.2   Distributed CMOS Amplifier

A different approach in the RF amplifier circuit design is presented in [11]. Due to the distributed topology, the gain of the amplifier is stable in a broad frequency range and therefore, the particular amplifier can be used in numerous applications. The inductors needed in this configuration are implemented by the package parasitics. The distributed amplifier uses the distributed capacitances of the MOS transistor channel in combination with the parasitic inductors of the packaging to create an artificial transmission line. This technique is extensively used in MMIC designs. The fact that the input impedance looking at the gate of a MOS transistor is capacitive, makes the CMOS tech-

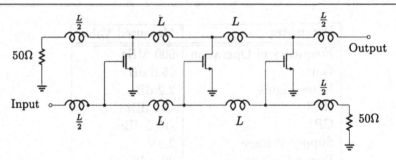

Figure 5.10: Distributed CMOS amplifier

nology an appropriate candidate for the implementation of distributed amplifiers. The combination of plastic packaging techniques along with mainstream CMOS technologies form a very attractive solution in terms of fabrication cost. Cable television (CATV) is a very important application in which the distributed amplifier technology can be applied. CATV calls for a constant gain in a very broad frequency range (50-850 MHz). In this case, the target is the RF front-end integration along with the signal processing components in a single chip solution.

The simplified schematic diagram of a three-stage distributed amplifier, is shown in Fig. 5.10.

Ignoring the losses in the circuit of Fig. 5.10, the small-signal gain of the amplifier is given by Eq. (5.14).

$$G = \frac{g_m^2 Z_{\pi g} Z_{\pi d}}{4} \left(\frac{\sin \frac{N}{2}(\beta g - \beta d)}{\sin \frac{1}{2}(\beta g - \beta d)} \right) \tag{5.14}$$

where $N = 3$ in the particular implementation, g_m is the transconductance of each stage, βg and βd are the phase constants of the gate and drain respectively and $Z_{\pi g}$ and $Z_{\pi d}$ are the impedances of the gate and drain respectively. If the phase constants become equal to each other, then the gain is frequency independent and it is maximized. Moreover, if the impedances become equal to Z_o, then the Eq. (5.14) becomes:

$$G = \left(\frac{N g_m Z_o}{2} \right)^2 \tag{5.15}$$

The noise factor of the amplifier is given by Eq. (5.16)

$$F = 1 + \frac{Z_o N \omega^2 C_{gs}^2 \delta}{3 g_m} + \frac{4\gamma}{N g_m Z_o} \tag{5.16}$$

Parameter	Measured Value
Bandwidth	300 kHz - 3 GHz
Gain	5 dB
Noise Figure @ 2 GHz	5.1 dB
IIP3 @ 2 GHz	+ 15 dBm
CP1 @ 2 GHz	+ 7 dBm
Input VSWR	1.7
Output VSWR	1.3
Supply Voltage	3 V
Current Dissipation	18 mA

Table 5.4: Performance of the amplifier of Fig. 5.10

where γ and δ are the noise factors of the drain and gate respectively for short-channel MOS devices, as mentioned in Chapter 2.

The first step towards the successful design of a distributed amplifier, is the proper selection and design of the waveguide. The selection of the L and C components of the waveguide define the impedance of the transmission line as well as the cutoff frequency of the waveguide. The cutoff frequency is given by Eq. (5.17):

$$f_c = \frac{1}{2\pi\sqrt{L(C_{gs} + C_p)}}$$ (5.17)

The gate-drain capacitance C_{gs} of the MOS transistor depends on the fabrication technology as well as on the geometry of the device. C_p is the total parasitic capacitance. In [11], the inductors have been formed by the package parasitics and therefore, there is not much freedom in the selection of their values. An alternative option could be the usage of on-chip inductors.

A three-stage distributed amplifier has been fabricated in a 0.8 μm CMOS technology [11] and its measurement results are reported in Table 5.4.

5.3.3   Tuned CMOS Amplifiers

Tuned amplifiers form a separate class of amplification stages. The implementation of such cells demands - apart from the utilization of high-speed transistor devices - the existence of passive devices (inductors, capacitors) of high quality. As already reported in Chapter 3, a lot of effort has been devoted in the modeling and characterization of Si integrated spiral inductors and this is a valuable aid to the tuned amplifier designer.

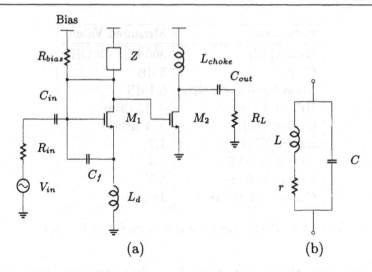

Figure 5.11: (a) Simplified circuit schematic of a tuned LNA (b) Simplified load Z

Modern RF subsystems design of a telecommunications transceiver, call for more than one integrated inductors on a common Si substrate [12-14]. In this case, the magnetic coupling between the various spiral inductors and transformers can drastically affect the performance of the system. Thus, the integrated circuit layout is directly associated to its electrical behavior.

A simplified schematic diagram of a tuned LNA using integrated inductors, is shown in Fig. 5.11. A load of impedance Z is employed in order to achieve tuning, as shown in Fig. 5.11(b). As shown in the figure, the load comprises a LC tank. The inductor exhibits ohmic losses that are denoted by the series resistor r. In practice, there is no need for a separate capacitor in order to achieve tuning: the parasitic capacitances of the integrated inductor to the substrate can be employed and the on-chip device can be designed in such a way that its first self-resonance frequency coincides with the wanted frequency of operation for the amplifier.

Using the simplified circuit of Fig. 5.11(b) for the load Z, the gain of the amplifier is given by Eq. (5.18):

$$\text{gain} = \frac{g_m + jQg_m}{r\left[g_d \cdot r - \omega^2 LC + 1\right) + j\omega(rC + g_d L)\right]} \tag{5.18}$$

where $Q = \omega L/r$ is the inductor's quality factor, g_m is the transconductance of M_1 and g_d is the drain conductance of M_1. It is evident from the above equation that the amplifier's gain is dominated by the quality of the integrated inductor used as a load. In Fig. 5.12, the simulation results of an integrated

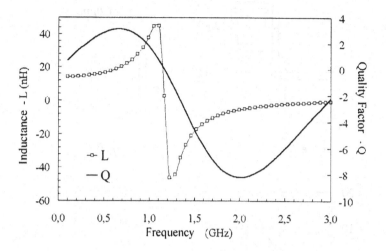

Figure 5.12: Inductance of load Z

square spiral inductor of 6 turns, outer dimensions of $470x470\mu m^2$ and width segment of 25 μm are shown. This particular inductor was used as a load Z. From the above figure, it is obvious that the self-resonance frequency of the inductor is around 1 GHz.

The second integrated inductor of the circuit is connected to the source of the device M_1 and is used to achieve input matching. In this case, the input impedance of the LNA is given by Eq. (5.19):

$$Z_{in} = \frac{1}{sC_{gst}} + sL_d + \frac{g_m L_d}{C_{gst}} \tag{5.19}$$

where $C_{gst} = C_{gs}||C_f$. C_{gs} is the small-signal capacitance between source and gate. From Eq. (5.19) is evident that by properly combining the values of the elements C_{gst} and L_d, the phase components cancel out at a particular frequency and thus, the circuit exhibits pure ohmic behavior at the input. By properly selecting the transconductance value g_m, matching at the required impedance level is also achieved.

In this particular design, the selected inductor is three-turn square spiral with outer dimensions of $470\mu mx470\mu m$.

Finally, the transistor M_2 forms an open drain configuration and is used to drive the proper load. As reported earlier, the magnetic coupling between the two inductors can seriously deteriorate the performance of the LNA. For example, a designer might choose to place the two integrated inductors the one on top of the other using different metalization layers in order to save silicon

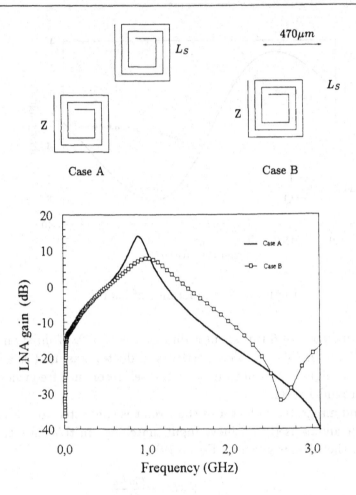

Figure 5.13: Effect of inductor placement on the LNA performance

area. Such a choice, drastically affects the circuit performance as shown in Fig. 5.13. In every case, the designer has to compromise between magnetic isolation between the spiral inductors and silicon area occupation. In that sense, digital CMOS technologies employing many metalization layers, favor the design of integrated circuits employing more than one spiral inductors.

A variant of the above topology is presented in [14]. In this case, the amplifier is differential (fully balanced) and the two RF components exhibiting a phase difference of 180°, can come from a - usually external - balun. The circuit schematic is shown in Fig. 5.14. In this topology, the input impedance

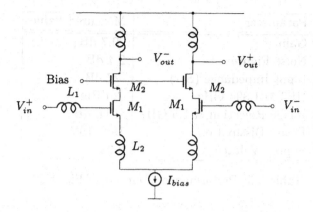

Figure 5.14: Differential tuned CMOS amplifier

is given by Eq. (5.20):

$$Z_{in} = s(L_1 + L_2) + \frac{1}{sC_{gs1}} + \frac{g_{m1}}{C_{gs1}} L_2 \qquad (5.20)$$

As usual, in order to achieve input matching, the following equations must hold:

$$(L_1 + L_2)C_{gs} = \frac{1}{\omega_0^2} \qquad (5.21a)$$

$$R_S = \frac{g_m L_2}{C_{gs}} \qquad (5.21b)$$

The simulation results for the above circuit, are summarized in Table 5.5. The tuned amplifier structure can overcome the inherent disadvantage of MOS-based designs which is the lower performance of the MOS transistor as opposed to that of the bipolar device (lower f_T and g_m/I_D). However, the overall performance of the tuned amplifiers is pertained by the quality of the integrated inductors used.

5.3.4 Tuned Bipolar Amplifiers

A tuned amplifier structure implemented in a bipolar process, can be found in [15]. It is a complete RF front-end design operating at 1.9 GHz in a silicon bipolar technology. The circuit schematic is shown in Fig. 5.15.

A common emitter configuration is used to implement the tuned amplifier. Tuning is achieved through the elements L_1 and C_1 at 1.9 GHz in this

Parameter	Measured Value
Gain	20.7 dB
Noise Figure	2.4 dB
Input Impedance (S_{11})	-27 dB
IIP3 @ 1.894 GHz	-2 dBm
Image Rejection @ 1.5 GHz	-10.6 dB
Power Dissipation	19.8 mW
Supply Voltage	3.3V

Table 5.5: Performance of circuit of Fig. 5.14

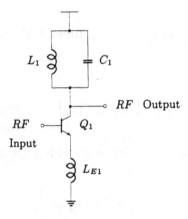

Figure 5.15: Tuned bipolar amplifier

case. The transistor Q_1 is biased by a 3 mA current and it is selected in order to optimize for noise performance. The feedback loop provided by the inductor L_{E1}, ensures better linearity and stability of the circuit along with the wanted input impedance matching. The LNA has been implemented in a $0.5\mu m$ bipolar technology with $f_T = 25$ GHz.

5.3.5   Tuned CMOS Amplifier with Active Q-Enhancement

The quality factor of a tuned amplifier is limited by the quality factor of the integrated inductor used as a load. A typical value for an integrated inductor's quality factor in a contemporary mainstream technology is 5. However, if an active Q-enhancement circuit is employed, very high values for the amplifier's quality factor can be achieved. It is therefore feasible to simultaneously meet two targets: the necessary input channel selection filtering along with the LNA circuit. In [16], such an implementation is presented in a $0.8\mu m$ CMOS

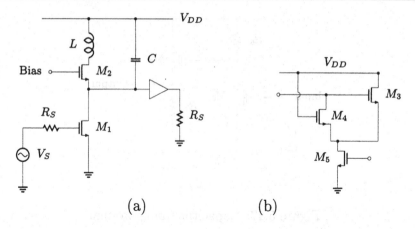

Figure 5.16: Tuned CMOS amplifier with active Q-enhancement

technology. The circuit schematic is shown in Fig. 5.16. In Fig. 5.16(a), the simplified circuit schematic is given and in Fig. 5.16(b), the active Q-enhancement topology is shown.

The Q-enhancement principle of operation is the formation of a negative conductance placed in parallel with the LC tuned circuit. This negative conductance cancels the ohmic losses of the integrated inductor [12]. If the negative conductance is G_n and the ohmic loss conductance of the integrated inductor is G_L, the total conductance of the configuration is $G_{TOT} = G_L - G_n$. The negative conductance G_n is achieved by the positive feedback circuit of Fig. 5.16(b). The transistor M3 acts as a source follower while the transistor M4 forms a common gate configuration. The active negative conductance is given by the following equation:

$$-G_n = -\frac{g_{m3}g_{m4}}{g_{m3} + g_{m4}} \qquad (5.22)$$

The conductances of transistors M3 and M4 are defined by the bias current of transistor M5. Thus, the total conductance $-G_n$ is tuned by the voltage V_{GS5}. This means that the total conductance G_{TOT} is tunable and therefore, the gain and quality factor of the amplifier are tunable.

The center frequency of the tuned amplifier is defined by the tank L_1C_1. It is imperative to employ an automatic frequency tuning scheme in order to absorb process tolerances. The proposed solution here is tuning through a Miller capacitance as shown in Fig. 5.17.

Due to Miller effect, the capacitance C_{eff} seen at the input of the amplifier is: $C_{eff} = (1 + A_f)C_f$ where A_f is the amplifier's gain given by the following equation: $-A_f = -g_{m6}R_6K_7$ where K_7 is the source follower gain. Gain

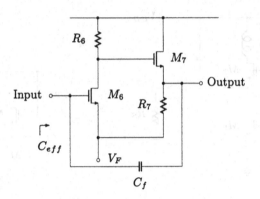

Figure 5.17: Capacitor tuning system

Parameter		Value
Supply Voltage		3V
Quality Factor		2.2 ÷ 44
$Q = 30$	Power Gain	17 dB
	Center Frequency	869 MHz - 893 MHz
	Noise Figure	6 dB
	IIP3	-14 dBm
	CP1	-30 dBm
	Power Consumption	78 mW

Table 5.6: Performance of the amplifier of Fig. 5.16

tuning is achieved through the transconductance g_{m6} which in turn is tunable through the voltage V_F applied at the source of the transistor.

Measurement results are summarized in Table 5.6. The high selectivity of the amplifier in combination with the center frequency tuning ability and the quality factor achieved, deteriorate the performance of the circuit as a LNA, mainly with respect to its noise figure and power consumption.

5.4 LOW NOISE AMPLIFIERS IN BiCMOS TECHNOLOGIES

Many of the topologies already presented have been also fabricated in BiCMOS technologies. This means that the bipolar devices that are used in this case usually exhibit half the speed of their pure bipolar transistor counterparts. A recent implementation for a WLAN transceiver at 2.5 GHz [17], contains a real BiCMOS LNA since it employs both MOS and bipolar transistors. The

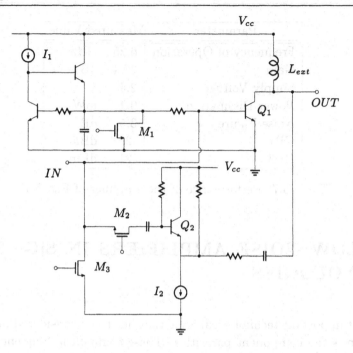

Figure 5.18: BiCMOS LNA

circuit schematic of the LNA is shown in Fig. 5.18.

The circuit operates in two modes: high gain and low gain state. The transistor Q_1 is the main amplification transistor in high gain operation. The bias of this circuit comes from a PTAT current produced by the bias current source I_1.

The transistor Q_2 in combination with the current source I_2 forms the low gain circuit. The transistor M_1 deactivates the high gain stage when it is on and the transistor M_2 connects Q_2 at the input of the amplifier so that M_3 is on and performs input impedance matching. In a similar manner, when the high gain mode is activated, the current source I_2 is off and thus, Q_2 is off. In this straightforward way, the switching between high and low gain states is performed.

The high gain value in this circuit is 14 dB and the low gain is -13 dB. The noise figure is 2.2 dB while CP1 is -15 dBm and IIP3 is -3 dB in high gain mode.

Parameter	Measured Value	
Frequency of Operation	6.25	GHz
Gain	20.4	dB
Supply Voltage	2.5	V
Power Dissipation	9.4	mW
Noise Figure	3.5	dB
CP1	-29	dBm
IIP3	-21	dBm

Table 5.7: Performance of the amplifier of Fig. 5.19

5.5   LOW NOISE AMPLIFIERS IN SiGe TECH-NOLOGIES

The most important technological issue that has to be considered in every RF IC design is the component parasitics. These confine the frequency range of operation of the active elements as well as the quality of the integrated passive devices which are equally important in a RF IC design.

Modern bipolar technologies invoke transistors with a cutoff frequency in the 30 - 50 GHz range. By employing SiGe devices, the limit of f_T has been pushed in the 50 - 80 GHz and, in some cases, in the 100 GHz range [18-20]. It is obvious therefore that the SiGe HBT technology is an ideal candidate for telecommunications applications and a serious competitor to the GaAs technology for a certain range of frequencies. Moreover, it is worth noting that the SiGe technologies retain the relatively low cost of the plain bipolar technology. Typical figures for passive devices reported in SiGe technologies, are as follows [18]:

Inductors in the range between a fraction of nH up to 100 nH exhibiting a quality factor between 3 and 20. High quality factor values are reported after modifying the substrate doping.

MOS capacitors exhibiting a quality factor around 20 / f(GHz) / C(pF).

MIM (metal-insulator-metal) capacitors exhibiting a quality factor around 80 / f(GHz) / C(pF).

Varactors with 40% tuning range and a quality factor around 70 / f(GHz) / C(pF).

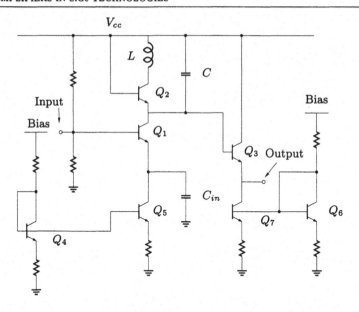

Figure 5.19: SiGe LNA at 6.25 GHz

5.5.1 Low Noise Amplifier at 6.25 GHz in a SiGe Technology

A recent LNA design in a SiGe technology (200 mm UHV/CVD Si/SiGe HBT from IBM) is reported in [21]. The particular technology exhibits an f_T value of 45 GHz. The circuit schematic is shown in Fig. 5.19.

The amplifier comprises a cascode stage followed by an emitter follower. The common base transistor Q_2 is used to increase the active impedance of the LC load and thus, the gain of the amplifier. It is also used (like in other topologies) to increase the isolation between input and output stages by reducing the Miller capacitance between the base and the collector of transistor Q_1. The input impedance is given by the following equation:

$$Z_{in} = \frac{1}{j\omega C_{\pi 1}}\left(1 + \frac{g_{m1}}{j\omega C_{in}}\right) \tag{5.23}$$

The inductor L is an on-chip inductor while the capacitors are of the MIM type. The measurement results of this LNA are summarized in Table 5.7.

Bibliography

[1] L.E. Larson, "Integrated Circuit Technology Options for RF IC's - Present Status and Future Directions," *IEEE 1997 CICC, paper no. 9.1.1*, pp. 169–176.

[2] R.G. Meyer and W.D Mack, "A 1-GHz BiCMOS RF Front-End IC," *IEEE Journal of Solid-State Circuits*, vol. 29, no. 3, pp. 350–355, March 1994.

[3] J.R. Long and M.A. Copeland, "A 1.9 GHz Low-Voltage Silicon Bipolar Receiver Front-End for Wireless Personal Communication Systems," *IEEE Journal of Solid-State Circuits*, vol. 30, no. 12, pp. 1438–1448, December 1995.

[4] S. Iversen, "The effect of feedback on noise figure," in *Proc. IEEE*, Mar. 1975, vol. 63, pp. 540–542.

[5] H. Fukui, "The noise performance of microwave transistors," *IEEE Trans. Electron Devices*, vol. 63, pp. 329–341, March 1966.

[6] G. C. Dawe, J. M. Mouraut, and A. P. Brokaw, "A 2.7V DECT RF Tranceiver with integrated VCO," *IEEE ISSCC 97, Session 18, Paper SA 18.5*, pp. 308–309.

[7] R. G. Meyer and R. A. Blauschild, "A 4 - Terminal Wide - Band Mondithic Amplifier," *IEEE J. Solid-State Circuits*, vol. SC-16, no. 6, pp. 634–638, Dec. 1981.

[8] K. H. Chan and R. G. Meyer, "A low - distortion monolithic wideband amplifier," *IEEE J. Solid-State Circuits*, vol. SC-12, pp. 685–690, Dec. 1977.

[9] A. N. Karanicolas, "A 2.7V 900 MHz CMOS LNA and Mixer," *IEEE J. Solid-State Circuits*, vol. 31, no. 12, pp. 1939–1944, Dec. 1996.

[10] A. N. Karanicolas, "A 2.7V 900 MHz CMOS LNA and Mixer," *IEEE ISSCC 96, Session 3, Paper TP 3.2*, pp. 50–51.

[11] P. J. Sullivan, B. A. Xavier, and W. H. Ku, "An Integrated CMOS Distributed Amplifier Utilizing Packaging Inductance," *IEEE Trans. on Microwve Theory and Techniques*, vol. 45, no. 10, pp. 1969–1976, October 1997.

[12] S. Pipilos, Y. Tsividis, J. Fenk, and Y. Papananos, "A Si 1.8 GHz RLC Filter with Tunable Center Frequency and Quality Factor," *IEEE J. Solid-State Circuits*, vol. 31, pp. 1517–1525, Oct. 1996.

[13] K. L. Fong, C. D. Hull, and R. G. Meyer, "A Class AB Monolithic Mixer for 900 MHz Applications," *IEEE J. Solid-State Circuits*, vol. 32, pp. 1166–1172, Aug. 1997.

[14] J. C. Rudell, J-J On, T. B. Cho, G. Chien, F. Brianti, J. A. Weldon, and P. R. Gray, "A 1.9 GHz Wide - Band IF Double Conversion CMOS Integrated Receiver for Cordless Telephone Applications," *IEEE ISSCC 97, Session 18, Paper 18.3*, pp. 304–305.

[15] J. Macedo, M. Copeland, and P. Schvan, "A 1.9 GHz Silicon Receiver with On - chip Image Filtering," *IEEE 1997 CICC*, pp. 181–184.

[16] C-Y Wu and S-Y Hsiao, "The Design of a 3-V 900 MHz CMOS Bandpass Amplifier," *IEEE J. Solid-State Circuits*, vol. 32, no. 2, pp. 159–168, Feb. 1997.

[17] R. G. Meyer and W. D. Mack, "A 2.5 GHz BiCMOS Transceiver for Wireless LAN," *IEEE ISSCC97, Paper 18.6*, pp. 310–311.

[18] J N. Burghartz, M. Soyuer, K. A. Jenkins, M. Kies, M. Dolan, K. J. Stein, J. Malinowski, and D. L. Harame, "Integrated RF Components in a SiGe Bipolar Technology," *IEEE J. Solid-State Circuits*, vol. 32, no. 9, pp. 1440–1445, Sept. 1997.

[19] A. Schüppen, H. Dietrich, S. Gerlach, H. Hohnemann, J. Arndt, U. Seiler, R. Gotzfried, U. Erben, and H. Schumacher, "SiGe Technology and Components for Mobile Communication Systems," in *Proc. Bipolar Circuits and Technology Meeting*, 1996, pp. 130–133.

[20] A. Schüppen, A. Gruhle, U. Erben, H. Kibbel, and U. Konig, "Multi - emitter finger SiGe HBT's with Fmax up to 120 GHz," in *Tech Dig. Int. Electron Devices Meeting*, 1994, pp. 337–380.

[21] H. Ainspan, M. Soyuer, J. O. Plouchart, and J. Burghartz, "A 6.25 GHz Low DC Power Low - Noise Amplifier in SiGe," *IEEE CICC 1997, Paper 9.2.l*, pp. 127–180.

CHAPTER
6

MIXERS

6.1 INTRODUCTION

Every modern telecommunications system uses at least one mixer, the performance of which largely determines the overall system performance. The term "mixer", in use since the earlier days when the super-heterodyne technique was invented, is inappropriate since a mixer does not actually "mix" the signals at its inputs (it does not add them linearly) but it multiplies them in pairs. Performing the task of driving the desired signal to the IF stage, out of a host of possible combinations of signals appearing at the antenna, is largely a matter of proper system design.

At the receiver, the mixer's role is to convert a high frequency (RF) signal to a more manageable low frequency (IF) signal (downconversion). At the transmitter, the mixer converts a low frequency signal to a high frequency one (upconversion), suitable for transmission after amplification. Any non-linear element can be used as a mixer: An inherent non-linearity is necessary for producing frequencies that do not exist at the input. Therefore, in mixer design several techniques are employed, according to the specific requirements of each application. Furthermore, the systems that perform signal multiplication can be divided in the following categories: multipliers, modulators and mixers.

Multiplier circuits have two input signals and they produce an output signal that is the linear product of the inputs. They fall into three sub-

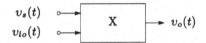

Figure 6.1: Multiplier circuit

categories: single-quadrant multipliers, when the input signals have only one polarity, two-quadrant multipliers, when one of the signals may change in polarity, and four-quadrant multipliers, when both input signals are free to take positive and negative values.

Modulators are sign modifying circuits, often called balanced modulators or doubly balanced modulators. They take two input signals and the output is one of the inputs multiplied by the other signal's sign.

Mixers are multiplier circuits, specially designed to optimally perform frequency translation. A mixer should exhibit excellent linearity as well as low noise. Mixers are either passive or active. The latter usually have a higher conversion gain.

Another categorization of mixer circuits is derived from the way that the output signal is created. This leads to multiplier-based mixers and switching mixers. In Fig. 6.1, a multiplier-based mixer is displayed. If $v_s(t) = V_s \cos(\omega_s t)$ and $v_{l_0}(t) = V_{l_0} \cos(\omega_{l_0} t)$ then $v_o(t) = \frac{K}{2} V_s V_{l_0} \left[\cos(\omega_s - \omega_{l_0})t + \cos(\omega_s + \omega_{l_0})t \right]^2$ where K is the multiplication gain.

The output will therefore contain a low frequency (IF) and a high frequency (RF) component. Consequently, an analog filter selects the appropriate signal: at the transmitter (upconversion) the RF component is selected, while at the receiver (downconversion) the IF component is selected.

In switching mixers, the LO signal controls the switches and thus the RF signal at the output appears modulated by a square wave. The desired frequency is selected by filtering. The operation of these circuits will be described further on. A good presentation of basic mixer topologies can be found in [1], where bipolar as well as MOS transistor circuits are analyzed, including the all-time classic Gilbert cell - the basis of many of the integrated circuit designs given further in this text.

6.2   BIPOLAR TRANSISTOR MIXERS

Before presenting certain integrated mixer implementations, it would be helpful to briefly examine the operation of the Gilbert cell, which is the most widely used topology in mixer design.

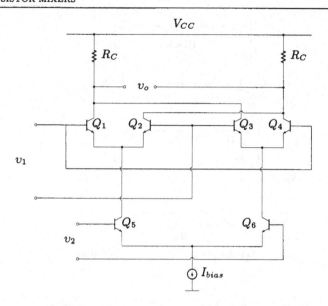

Figure 6.2: Four-quadrant multiplier

A fully balanced, four-quadrant multiplier is shown in Fig. 6.2. The differential output voltage of this circuit is given by the following expression:

$$
\begin{aligned}
v_o &= -R_c[(I_{c1} + I_{c3}) - (I_{c2} + I_{c4})] = \\
&= -R_c[(I_{c1} - I_{c2}) + (I_{c3} - I_{c4})] = \\
&= -R_c\left[(I_{c5} - I_{c6})\tanh\left(\frac{v_1}{2V_T}\right)\right] = \\
&= -R_c I_{bias}\tanh\left(\frac{v_2}{2V_T}\right)\tanh\left(\frac{v_1}{2V_T}\right)
\end{aligned}
\tag{6.1}
$$

Therefore, for small input voltage signals the output becomes:

$$
v_o = -\frac{R_c I_{bias}}{4V_T{}^2}v_1 v_2
\tag{6.2}
$$

In this fashion, multiplication of signals v_1 and v_2 is effected. The linearity issue that was previously identified as a major mixer design concern, needs to be resolved. Towards increasing the linear operating range of the multiplier, Barrie Gilbert invented a scheme for pre-distorting the input signal. This is achieved by the use of a pair of diodes connected at the input of an emitter-coupled pair (ECP), which compresses the input in a logarithmic law.

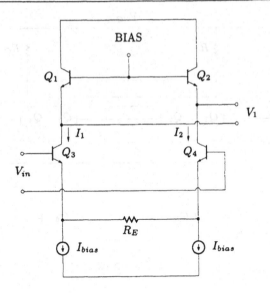

Figure 6.3: Voltage-to-current converter

In Fig. 6.3, a voltage-to-current converter used as a means of signal compression is depicted. Resistor R_E is employed for ECP linearization. The differential output current is :

$$I_1 - I_2 = \frac{2v_{in}}{R_E} \qquad (6.3)$$

The converter's output becomes the input to transistors $Q_1 - Q_4$ in the circuit of Fig. 6.2, comprising a Gilbert cell. This results in a linearized mixer.

6.2.1   Class AB Mixer at 2.4 GHz

Recently [2] a class AB mixer was proposed for wireless network applications. This specific topology achieves low noise level and high linearity, which are the most important features of a mixer.

The topology is displayed in Fig. 6.4, where a buffer circuit converts the single-ended LO signal to differential (Fig. 6.5).

The mixer comprises two parts : an emitter follower (Q_1) and a switch pair (Q_2, Q_3). The amplifier (Q_1) is employed for driving the RF signal to the mixer and compensating for the switching losses in transistors Q_2 and Q_3. The latter perform the signal mixing and produce a differential IF.

In a similar case to the one presented in Chapter 5, inductance L_e (belonging to a bonding wire) is applied to increase the linearity of the common emit-

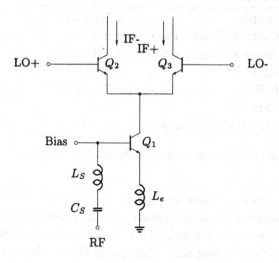

Figure 6.4: Class AB mixer

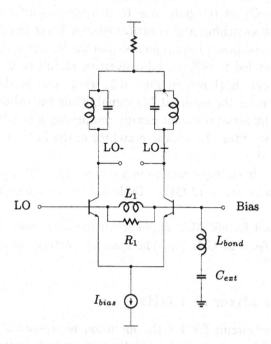

Figure 6.5: Single-ended to differential converter

Parameter	Measured Value
Supply	$2.7 \div 5.5V$
Current Dissipation	8 mA
Power Conversion Gain	4.5 dB
IIP3	1 dBm
ICP1	-7.5 dBm
SSB NF	10 dB
LO - RF Feedthrough	-28 dBc

Table 6.1: Performance of the mixer of Fig. 6.4

ter stage and does not contribute any noise - as would be the case if an ohmic resistor was used instead for degeneration. On the other hand, inductance L_e should be adequately small to maximize the common emitter stage gain, thus justifying the use of a bonding wire. Transistor Q_1 should be designed large enough to minimize its r_b and therefore noise. Also, an appropriate choice for L_e and Q_1 yields 50Ω matching at the mixer's input.

The lack of resistive feedback ensures a class AB operation, leading to increased linearity for a given bias current. Non-linear behaviour is mainly caused by the Q_2-Q_3 switch pair, due to non-ideal switching. A large LO signal enables fast switching and contributes to a lower circuit noise profile. To alleviate the signal power requirements from the (external) VCO, the mixer employs a single-ended to differential converting circuit to drive the bases of Q_2, Q_3. The circuit is shown in Fig. 6.5, where one of the transistors of the input pair receives the external LO signal, while the other is driven to ac ground through the series resonant circuit comprising a bondwire inductance and an external capacitor. To achieve matching at the LO input, an integrated inductor L_1 is used.

The circuit has been implemented in a $0.8\mu m$ BiCMOS process, where the f_T of the npn transistor is 13 GHz. Table 6.1 summarizes the measurement results.

A similar circuit for 900 MHz applications has been fabricated in a bipolar technology with $f_T = 25$ GHz [7]. The basic characteristics of this mixer are given in Table 6.2.

6.2.2 Bipolar Mixer at 1 GHz

A switching mixer circuit for 1 GHz operation is presented in [3]. The LO signal is produced by an external oscillator and controls switches Q_1 and Q_2, as depicted in Fig. 6.6. The RF signal is introduced through transistor Q_3,

Parameter	Measured Value
Supply	$2.7 \div 5V$
Current Dissipation	10.1 mA
LO Signal Power	-10 dBm
Power Conversion Gain	7.5 dB
IIP3	2.5 dBm
ICP1	-1.5 dBm
SSB NF	12.1 dB
LO - RF Feedthrough	-47.5 dBc

Table 6.2: Performance of a class AB mixer at 900 MHz

Parameter	Measured Value
Supply	5 V
Voltage Gain	10 dB
IIP3	$+6$ dBm
ICP1	-4 dBm
NF	15.8 dB
LO - RF Feedthrough	-46 dB

Table 6.3: Performance of the mixer of Fig. 6.6

which forms a common base buffer. Resistor R_1, combined with the input resistance of Q_3 at the base, $\left(\frac{1}{g_m}\right)$, provide input matching for the mixer at high frequencies. The 50Ω resistance of the RF source along with R_1, result in excellent linearity through emitter degeneration. Due to the operation of switches Q_1 and Q_2, the IF component of the output current is described by the Fourier transform:

$$ I_{IF} = \frac{V_{RF}}{100} \frac{1}{\pi} \cos(\omega_{LO} - \omega_{RF})t \qquad (6.4) $$

This component should be selected amid other components of the Fourier series, by use of an external filter.

The above circuit has been implemented in a $0.8\mu m$ BiCMOS process with $f_T = 13$ GHz for the npn transistor. The measurement results from the overall mixer circuit are summarized in Table 6.3. It is noted that the required LO power for switching is 0 dBm.

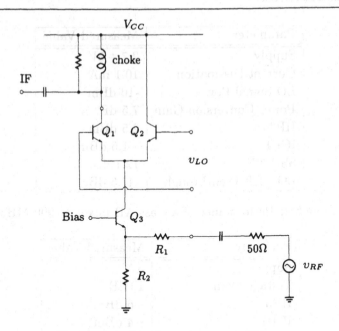

Figure 6.6: Bipolar mixer at 1 GHz

6.2.3   Doubly Balanced Mixer

Balanced mixers generally outperform their single-ended counterparts [4]. In Fig. 6.7, the schematic of a balanced mixer for 1.9 GHz applications is shown [5].

As the figure shows, balun T_1 converts the RF input signal to RF_+ and RF_-. These two signals consequently drive the four transistor switches $Q_1 - Q_4$. The bias current for the latter is supplied by Q_5, via the center tap of balun T_1. The switching operation of $Q_1 - Q_4$ produces the desired output IF signal that can drive the next stage through an external Darlington buffer or an external matching circuit (not shown in Fig. 6.7).

The balanced topology of the above mixer enables the cancellation of all unwanted even order harmonics at the IF output. For proper switching, a strong LO signal is required, along with a high isolation between the LO port and the RF and IF ports of the mixer. Using the balun T_1 yields very good isolation between LO and RF.

The circuit has been implemented in a $0.8 \mu m$ BiCMOS process with $f_T = 11$ GHz for the npn transistor. Table 6.4 summarizes the mixer's performance.

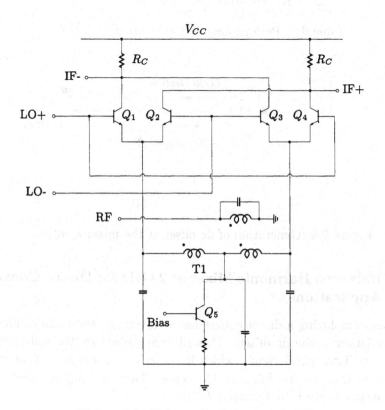

Figure 6.7: Balanced mixer at 1.9 GHz

Parameter	Measured Value
Supply	1.9V
Current Dissipation	2.5 mA
Conversion Gain	6.1 dB
Operating Frequency	1.9 GHz
IIP3	2.3 dBm
SSB NF	10.9 dB
LO - RF Feedthrough	-32 dB

Table 6.4: Performance of the mixer of Fig. 6.7

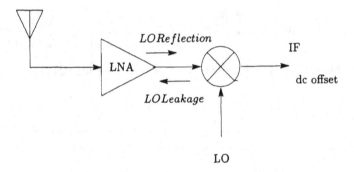

Figure 6.8: Generation of dc offset at the mixer's output

6.2.4   Balanced Harmonic Mixer at 2 GHz for Direct-Conversion Applications

In receivers employing a direct conversion architecture, the main challenge is the cancellation of the dc offset. This effect is caused by the self-mixing of the local oscillator (LO) signal, which leads to a dc component that cannot be discerned from the useful, zero-IF signal. The resulting dc offset at the mixer's output is given by Equation (6.5) :

$$V_{off} = V_{leak} R_{amp} G_{LO} \qquad (6.5)$$

where V_{leak} is the leakage voltage of the LO signal at the RF port of the mixer, R_{amp} is the reflection coefficient of the LNA-antenna system that precedes the mixer and defines the LO leakage amount appearing at the RF port, and finally, G_{LO} is the $LO - IF$ conversion gain. The LO self-mixing procedure is depicted in Fig. 6.8.

Therefore, the mixer's characteristics that need to be minimized in order

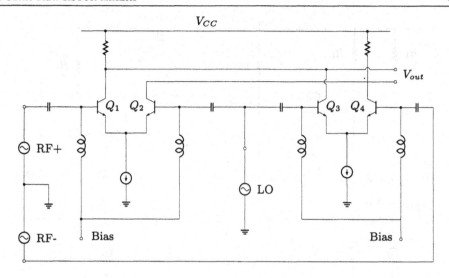

Figure 6.9: Balanced harmonic mixer

Parameter	Measured Value
Supply	2.7 V
Current Dissipation	5 mA
Conversion Gain	5.6 dB
IIP3	−1 dBm
NF	17 dB
dc offset	< −92 dBm

Table 6.5: Performance of the mixer of Fig. 6.9

to reduce the *dc* offset at the output, are the $LO - RF$ leakage and the LO conversion gain.

In the circuit of Fig. 6.9 a balanced harmonic mixer (BHM) is presented, aimed at direct conversion systems and achieving *dc* offset minimization [6]. Due to the symmetry in the circuit, if the $Q_1 - Q_2$ and $Q_3 - Q_4$ pairs are well matched, LO leakage at the RF port is eliminated. Transistors $Q_1 - Q_4$ operate as switches, as in the previous cases. The circuit has been implemented in a bipolar technology and the experimental results are shown in Table 6.5, for a LO signal power of $-6dBm$.

6.2.5   Gilbert Mixer at 900 MHz with 1.5V Supply

In Fig. 6.3, the conventional voltage-to-current converter was introduced, where the R_E resistor provided emitter degeneration for better linearity. The

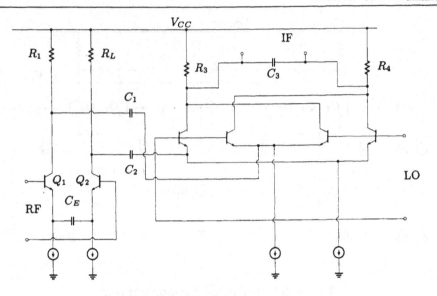

Figure 6.10: Gilbert mixer with capacitive linearization

main drawbacks of this technique are : (a) the thermal noise of R_E raises the overall circuit noise, and (b) a voltage supply higher than $2V_{BE}$ Volts(Q_1, Q_3) should be used. The first drawback can be eliminated if a capacitor is used in the place of R_E, with an appropriate capacitance value for the frequency range of interest. Also, the supply voltage can be reduced if ohmic resistors are used as loads in the place of Q_1 and Q_2, provided that the voltage drop across each resistor is lower than V_{BE}.

A mixer with the above improvements is presented in Fig. 6.10. It was designed [8] for 900 MHz operation, using a $1\mu m$ BiCMOS process where the $f_T = 20$ GHz for the npn transistor. The measurement results of this mixer are given in Table 6.6.

In this design an effort was made to reduce noise : Firstly, the current sources were implemented with MOS transistors that produce three times lower noise than the respective bipolar ones. Thus, a BiCMOS design was attained. Moreover, the coupling capacitors C_1 and C_2 were given such values so that their bottom plate parasitics would yield low-pass filters with cutoff frequencies near 1 GHz (mixer's frequency of operation). As a result, the high frequency noise components derived by LO harmonics have lower contribution to overall noise.

Parameter	Measured Value
Supply	1.5 V
Power Dissipation	15 mW
Conversion Gain	4 dB
IIP3	3 dBm
CP1	−10 dBm
NF	15 dB
LO - RF Feedthrough	-37 dB

Table 6.6: Performance of the mixer of Fig. 6.10

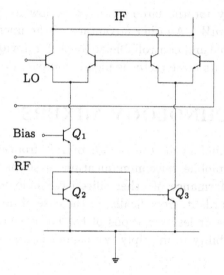

Figure 6.11: Micromixer schematic

6.2.6 Micromixer

Recently [9,10], Barrie Gilbert introduced an alternative approach to mixer linearization and operation. The micromixer technique was originally tried by Analog Devices in 1994, while in 1997 [10] a DECT transceiver employing this technique at the receiver part was implemented in a bipolar process with $f_T = 25$ GHz.

The micromixer circuit is thoroughly analyzed in [11]. The core of the circuit that effects mixing is the same with that of the conventional Gilbert cell. The sole difference is at the *RF* port and is attributed to the way that large input signals are handled. The basic micromixer schematic is shown in Fig. 6.11. The micromixer operates in class AB.

Transistor Q_1 forms a common base configuration and conveys its output

- the collector current of Q_1 - to the base of the mixing cell with zero phase delay. The current mirror Q_2/Q_3 carries its (I_{CQ3}) output to the mixer with a 180° phase delay. in theory, this topology can handle arbitrarily high currents with increased linearity.

Currents I_{CQ1} and I_{CQ3}, with reference to v_{RF}, are generally linear, but for large input signals I_{CQ1} is linear for negative RF voltages while I_{CQ3} is linear for positive RF voltages. Since the system operates in class AB, only one output is valid per half period. The sum of the two signals at the IF output results in a linear mixer behaviour all over the period. This technique can yield $IP3$ values up to $+30dBm$ with relatively low power consumption.

The voltage supply for the circuit can be as low as $2.2V$ with a total dissipation less than $5mW$. Another advantage of the micromixer is that its input resistance is stable and controllable at (50Ω) for a wide frequency band. A disadvantage is that its noise factor is higher compared to other topologies.

6.3   MOS TECHNOLOGY MIXERS

The use of MOS technology in the design of RF front-ends constitutes a low-cost solution for mobile telecommunications systems. Especially since MOS reaches the performance of other silicon technologies, the industry is gradually turning towards it. Specifically, in the case of mixers, the proposed MOS circuits are more or less variations of the Gilbert cell, while the MOS transistor's inherent ability to multiply two signals is exploited.

6.3.1   Active Variable Gain Mixer

The circuit of a balanced MOS mixer used in an image-rejecting system for DECT applications, is shown in Fig. 6.12 [12].

Transistors $M5 - M8$ form the Gilbert cell core where the LO signal is injected. Transistors $M1$ and $M2$ are used at the RF input, while $M3$ and $M4$ are used in a cascode configuration, in order to increase the isolation between LO and RF/IF.

Transistors $M9$ and $M10$ operate in the triode region and their resistance is therefore defined by their gate voltage, which subsequently depends on the current I_G flowing through the diode-connected transistor $M16$. The mixer's load and gain are thus determined. The conversion gain can be varied from 0 to $10dB$.

The current sources $M11$ and $M12$, along with the common-mode feedback loop $M13 - M15$, adjust the common-mode signal level at the output. The

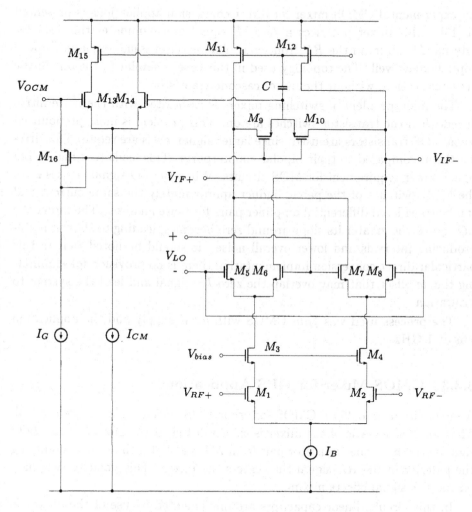

Figure 6.12: Active variable gain mixer schematic

design has been implemented in a $0.6\mu m$ analog CMOS process, with a $3V$ supply.

6.3.2 Direct Conversion Mixer at 1 GHz

An experimental CMOS mixer for direct conversion applications is presented in [13]. This mixer produces a zero-IF signal and eliminates the need for external IF filters in the RF front-end. The dc offset issues explained above apply here as well. The topology used in this case is similar to the one in the previous section, without the use of cascode transistors.

The main problem of switching mixers is associated with the noise introduced when the transistor switches are on. This problem is more pronounced when MOS transistors are used, since larger signal levels are required for driving them, compared to their bipolar counterparts. This leads to a higher LO signal power requirement for MOS circuits. When the LO signal crosses zero, the MOS switches of the pairs conduct approximately the same current and can be considered differential amplifier pairs for noise analysis. The larger the LO signal, the greater its slope around zero becomes, leading to shorter noise-producing intervals and lower overall noise. It should be noted that in this particular direct conversion implementation there is no provision for eliminating the dc offset that may overlap the zero-IF signal and lead the system to saturation.

The process used was $1\mu m$ CMOS with a $3V$ supply and the application was at 1 GHz.

6.3.3 CMOS Mixer for GPS Applications

A mixer circuit in $0.35\mu m$ CMOS for use in GPS systems, is presented in [14]. The circuit schematic of the mixer is shown in Fig. 6.13. The LO_+ and LO_- signals switch on one transistor pair from $M1 - M4$ at a time, thus modifying the polarity of the RF signal that enters the mixer. This polarity switching on the RF signal effects mixing.

In this circuit, linear capacitors are implemented by use of the dielectric layer found between two different layers of metallization. In a submicron CMOS process the resulting capacitance per unit area is fairly high and this specific capacitor structure is preferred over the non-linear MOS parasitic capacitances.

Inductors are implemented with bonding wires to yield an adequately high Q. The measurement results of the above mixer are summarized in Table 6.7.

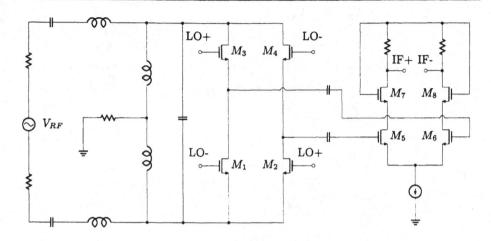

Figure 6.13: Mixer for GPS applications

Parameter	Measured Value
LO Frequency	1.4 GHz
LO Amplitude	300 mV
Supply	1.5 V
Voltage Conversion Gain	−3.6 dB
IIP3	10 dBm
CP1	−5 dBm
NF (SSB)	10 dB

Table 6.7: Performance of the mixer of Fig. 6.13

6.3.4 Broadband Mixer with a Multiplication Method

A different mixer topology employing MOS transistors in triode that modifies the signal according to a voltage applied at the gate, appears in [15]. It was used to design a doubly balanced zero-IF $I - Q$ modulator at 900 MHz [16]. A similar high-linearity technique is also found in [17]; the latter is however based on sampling schemes.

The basic concept of the linear CMOS mixer utilizing multiplication methods, is depicted in Fig. 6.14. The circuit is based on [18] and [30]. The main advantage of the proposed topology is its high linearity. The four MOS transistors in Fig. 6.14 achieve the desired linearity, since the fully balanced structure cancels out the non-linearities associated with the g_{ds} of the transistors. The RF signal at the transistors' gates modifies their conductance. The transfer function is given by Equation (6.6) :

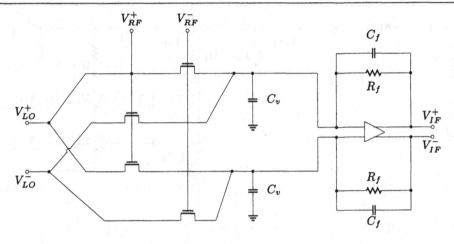

Figure 6.14: CMOS mixer with a multiplication method

$$(V_{IF}{}^{+} - V_{IF}{}^{-}) = \beta R_f (V_{RF}{}^{+} - V_{RF}{}^{-}) \cdot (V_{LO}{}^{+} - V_{LO}{}^{-}) \qquad (6.6)$$

In the above equation, β is the transconductance coefficient of the MOS. The requirements from the employed operational amplifier are relaxed, since it only processes the IF signal (up to a few MHz). For better results, capacitors C_v are included to drive the high-frequency products to the ground.

However, the above topology has several inherent problems: The output noise is higher than that of the Gilbert cell mixers and the conversion gain is significantly lower due to multiplication. The latter can be very annoying if the LO power supplied by the oscillator is low. Mismatches between MOS transistors are the main cause of distortion and give rise to the square of signal V_{LO} at the output. This produces an extra dc component which is unwanted and a component at two times the LO frequency which is filtered by the C_v capacitors.

Finally, another problem is caused by the parasitic capacitances of the transistors, that result in lower isolation between LO, RF and IF signals. To suppress this effect, the parasitics should be reduced by decreasing the size of the transistors. However, if the bias currents are kept the same, this will lead to lower transconductance and thus, lower conversion gain.

The circuit has been implemented in a $1.2\mu m$, 5V CMOS process. Power consumption is extremely low, since the multiplication method is employed. Measurement results are given in Table 6.8.

Parameter	Measured Value
Operating Frequency	up to 1.5 GHz
IIP3	45 dBm
NF	32 dB
Supply	5 V
Power Dissipation	1.3 mW

Table 6.8: Performance of the mixer of Fig. 6.14

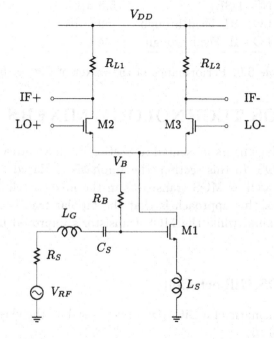

Figure 6.15: Mixer circuit with current reuse

6.3.5 NMOS Mixer for NADC Applications

A mixer system using NMOS transistors is shown in Fig. 6.15 [19-20]. Transistor $M1$ is used as a transconductance amplifier, while transistors $M2$ and $M3$ effect mixing through switching by the LO signal applied at the transistors' gates. Input matching is achieved by use of inductors L_G and L_S. The current reuse technique that was described for the LNA of Par. 5.2.1 is also employed by this circuit.

The above mixer was implemented in a $0.5\mu m$ CMOS process. The results are summarized in Table 6.9.

Parameter	Measured Value
Supply	2.7 V
Power Dissipation	7 mW
RF Frequency	900 MHz
LO Frequency	1 GHz 0 dBm
Power Conversion Gain	8.8 dB
IIP3	−4.1 dBm
CP1	−16.1 dBm
NF (DSB)	5.8 dB
LO - RF Feedthrough	-59.6 dB
LO - IF Feedthrough	-34.4 dB

Table 6.9: Performance of the mixer of Fig. 6.15

6.4   BiCMOS TECHNOLOGY MIXERS

Some of the bipolar mixers presented in Section 6.2 have already been implemented in BiCMOS. In this section, the emphasis is placed on circuits that utilize bipolar as well as MOS transistors in the mixer's cell. The most obvious advantage of this approach is that the bipolar transistor can be used for optimal switching, while the MOS transistor's improved linearity is also exploited.

6.4.1   BiCMOS Gilbert Cell

The simplified schematic of a BiCMOS mixer based on the classic Gilbert cell is shown in Fig. 6.16.

As the figure shows, the MOS transistors $M1$ and $M2$ are used at the mixer's RF input to ensure the best possible linearity, while the four bipolar transistors $Q_1 - Q_4$ perform mixing with the LO signal.

A thorough analysis of this mixer's operation, considering an ideal pulse at the bases of the bipolar transistors, yields :

$$i_A - i_B = i_1 - i_2 \qquad \text{if } v_{LO+} \text{ high} \tag{6.7}$$

$$i_A - i_B = i_2 - i_1 \qquad \text{if } v_{LO+} \text{ low} \tag{6.8}$$

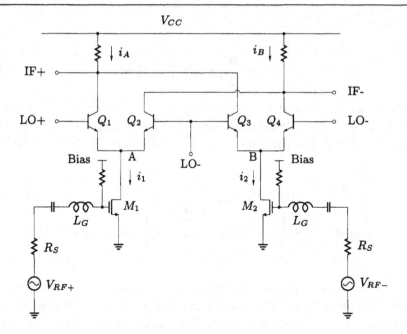

Figure 6.16: BiCMOS Gilbert mixer

Therefore, $i_A - i_B = (i_1 - i_2)p(t)$ where $p(t) = 1$ if v_{LO+} is high and $p(t) = -1$ if v_{LO+} is low. Signal $p(t)$ is a square pulse of LO frequency. A Fourier analysis of $p(t)$ gives :

$$i_A - i_B = (i_1 - i_2)[\frac{4}{\pi}\cos\omega_{LO}t + \frac{4}{3\pi}\cos 3\omega_{LO}t + \frac{4}{5\pi}\cos 5\omega_{LO}t + \cdots] \quad (6.9)$$

Currents i_1 and i_2 are the drain currents of the MOS transistors, which operate in saturation. Therefore :

$$i_1 = \frac{1}{2}\mu C_{ox}'\frac{W}{L}(v_{GS1} - V_T)^2 = \frac{1}{2}\mu C_{ox}'\frac{W}{L}(V_G + v_{RF+} - V_T)^2 \quad (6.10)$$

$$i_2 = \frac{1}{2}\mu C_{ox}'\frac{W}{L}(v_{GS2} - V_T)^2 = \frac{1}{2}\mu C_{ox}'\frac{W}{L}(V_G + v_{RF-} - V_T)^2 \quad (6.11)$$

Thus, the difference of the two currents becomes:

$$i_1 - i_2 = \frac{1}{2}\mu C_{ox}'\frac{W}{L}[(V_G + v_{RF+} - V_T)^2 - (V_G + v_{RF-} - V_T)^2] \quad (6.12)$$

$$i_1 - i_2 = \frac{W}{L}\mu C_{ox}{}'(V_G - V_T)v_{RF} \tag{6.13}$$

where v_{RF} is the total RF signal $(v_{RF+} - v_{RF-})$ and $v_{RF} = V_{RF}\cos\omega_{RF}t$. From Equation (6.13) it becomes obvious that, if the MOS transistors are perfectly matched, linearization is ideal. Replacing (6.13) in (6.9), yields :

$$i_A - i_B = \frac{W}{L}\mu C_{ox}{}'(V_G - V_T)V_{RF}\Big\{\frac{4}{\pi}\Big[\frac{1}{2}\cos(\omega_{LO} - \omega_{RF})t + \frac{1}{2}\cos(\omega_{LO} +$$
$$\omega_{RF})t\Big] + \frac{4}{3\pi}\Big[\frac{1}{2}\cos(3\omega_{LO} - \omega_{RF})t + \frac{1}{2}\cos(3\omega_{LO} + \omega_{RF})t\Big] + \cdots\Big\}$$
$$\tag{6.14}$$

From the above equation, the term $\omega_{LO} - \omega_{RF} = \omega_{IF}$ is the desired frequency. All other frequencies can be rejected by filters following the mixer. As a result, the IF component of the differential current $(i_A - i_B)$ will be :

$$i_A - i_B = \frac{W}{L}\mu C_{ox}{}'(V_G - V_T)V_{RF}\frac{4}{\pi}\frac{1}{2}\cos(\omega_{LO} - \omega_{RF})t \tag{6.15}$$

The amplitude of the IF current I_{IF} is :

$$I_{IF} = G_C \cdot V_{RF} \tag{6.16}$$

where G_C is the large signal transconductance gain :

$$G_C = \frac{2}{\pi}\mu C_{ox}{}'\frac{W}{L}(V_G - V_T) \tag{6.17}$$

In this particular topology, RF input matching is provided by inductances L_G that compensate for the phase shift caused by MOS parasitics. In the circuit, the common node of bipolar transistor emitters is the ac ground. The L_G elements may be implemented by the leads of the IC package and by PCB inductors, if necessary.

The LO signal is balanced $(LO+, LO-)$ and is derived by a single-to-differential (S-D) converter, also included in the integrated circuit. Therefore, the 50Ω matching issue for the LO input is shifted from the mixer to the converter.

Parameter	Simulated Value
Supply	2.7 V
Current dissipation	6 mA
LO signal power	−10 dBm
LO signal frequency	900 MHz
RF signal frequency	1 GHz
Power conversion gain	−1 dB
IIP3	+7.5 dBm
NF	10 dB

Table 6.10: Simulation results from the mixer in Fig. 6.16

The power conversion gain G_α of the mixer depends on the total current consumption I_{PS}, according to the following relation :

$$I_{PS} = \pi(V_G - V_T)\sqrt{\frac{G_\alpha}{R_o R_s}} \qquad (6.18)$$

where R_S is the RF source resistance and R_o is the load at the IF output. I_{PS} is the sum of the currents flowing through the drains of the two MOS transistors.

Thus, the conversion gain is related to the current consumption, which in turn is dependent on MOS geometry, considering that the $(V_{GS} - V_T)$ difference should be adequately high to achieve good linearity.

Table 6.10 summarizes the simulation results of a BiCMOS Gilbert mixer, in a $0.7\mu m$ BiCMOS process where the npn transistor's f_T is around 11 GHz.

A variation of the above mixer where the MOS transistors at the RF input are replaced by bipolar ones, is shown in Fig. 6.17 [21]. Linearization is effected by use of inductors L_E at the emitters of Q_1 and Q_2 (inductive degeneration). These inductors have a nominal value of $1nH$ and are implemented on chip. The use of these inductors instead of ohmic resistors enables linearization and input matching, without significantly raising the noise factor of the circuit.

This mixer was implemented in a $0.7\mu m$ BiCMOS process, where the f_T for the npn transistor was at 20 GHz. The measurement results are given in Table 6.11. The circuit was used in a wireless LAN (WLAN) transceiver system, operating at 2.5 GHz.

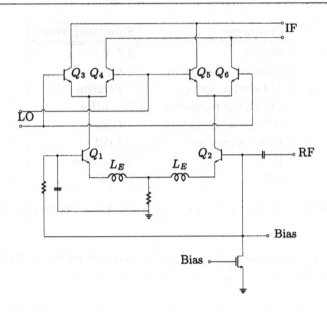

Figure 6.17: BiCMOS mixer for WLAN applications

Parameter	Measured Value
Supply	2.7 V
LO Signal Power	−13 dBm to +5 dBm
Conversion Gain	8 dB
IIP3	+3 dBm
SSB NF	11 dB
Operating Frequency	2.45 GHz

Table 6.11: Performance of the mixer of Fig. 6.17

6.5   SiGe TECHNOLOGY MIXERS

As it was mentioned in the previous chapter, the use of SiGe heterojunction
bipolar transistors (HBT) boosts integrated circuit design to higher frequencies. In [22], a receiver mixer operating in the vicinity of 2.5 GHz is presented,
based on [5]. The mixer's schematic is shown in Fig. 6.18. The coupling of the
RF signal is effected through a transformer. The latter is part of a negative
feedback loop, that offers the added benefit of lower distortion and increased
dynamic range for the circuit that operates from a mere $1V$ supply.

The input balanced-to-unbalanced transformer (balun) is fully integrated.
On the contrary, the balun for the IF output is located off-chip, to attain
higher flexibility in selecting the required IF. The RF input matching is near

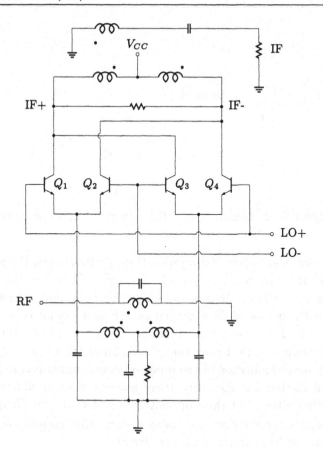

Figure 6.18: SiGe mixer schematic

50Ω, with low dependence on the supply current (approx. $2.5mA$).

The measured $IIP3$ value for the above mixer is $+4dBm$, for a LO signal frequency of 2.5 GHz.

6.6 IMAGE - REJECTION MIXERS (IRM)

6.6.1 The Image Problem

In a telecommunications system, besides the desired carrier frequency in each case, several other signals manifest themselves (e.g. adjacent channels, random interference) and should be rejected. One of the most detrimental is the image frequency signal (IM) : If a frequency ω_{RF} appears at the mixer input of a receiver, the resulting output signal after mixing with frequency ω_{LO} of the local oscillator will be $\omega_{RF} - \omega_{LO} = \omega_{IF}$. The image frequency's location in

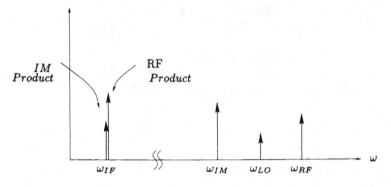

Figure 6.19: Generation of the unwanted image frequency

the band is such that, after mixing with the ω_{LO} frequency, the following ω_{IF} product appears (Fig. 6.19) : $\omega_{LO} - \omega_{IM} = \omega_{IF}$. Therefore, the image signal is located at $\omega_{RF} - 2\omega_{IF}$. It is obvious that the IM signal is unwanted, since it overlaps with the useful data carried by RF and cannot be extracted from it. To reject the IM, two architectures are usually applied : Image rejection by use of a bandpass filter before the mixer's input (this is usually an external component), and application of the image-rejection mixer techniques that will be presented further on. Recently, there have been some efforts to integrate image-rejecting filters, but their presentation is deferred for Chapter 8.

Image-rejection mixers are invariably systems that employ two simple mixers, two phase shifters and a combiner circuit.

6.6.2   Image - Rejection Mixer at 2.5 GHz

In [23] a bipolar image-rejection mixer is presented, implemented in a BiCMOS process where the f_T of the bipolar transistor is 15 GHz. The simplified circuit schematic is depicted in Fig. 6.20. The mixers are Gilbert cells, while the 90^0 phase shifters $(\Delta - \Phi)$ are RC/CR networks, as in Fig. 6.21.

The rationale behind using RC/CR networks as phase shifters in this topology, is to lower the power consumption. In microwave systems, transmission lines are usually employed, that can provide a 90^0 phase shift. However, in silicon technologies these structures are not feasible. From Fig. 6.21 it is derived that the role of the phase shifters is to discriminate between the phase of RF and the phase of IM products, so that they are ultimately added with opposite signs at the output combiner and are thus cancelled out. It is obviated that the image-rejection amount depends on the matching between the signals, and consequently on the close matching between the R and C

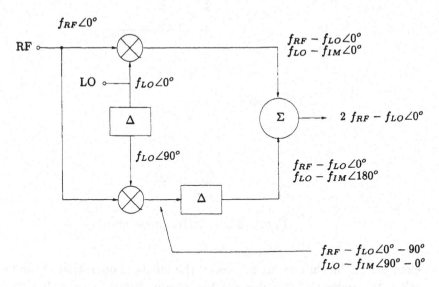

Figure 6.20: Image-rejection mixer schematic

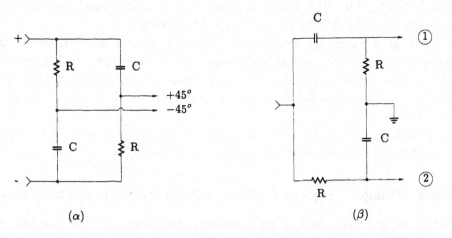

Figure 6.21: RC/CR phase shifters

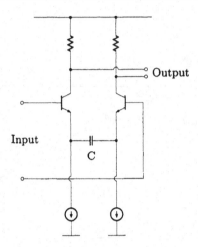

Figure 6.22: Active phase shifter

elements of the shifters, as well as on the identical operation of the two mixer cells. An analysis of the phase shifter of Fig. 6.21(b) shows that its two outputs (1 and 2) have a constant phase difference of 90^0 and an amplitude ratio of $\frac{A_1}{A_2} = 2\pi f RC$. Since the two amplitudes should ideally be equal for any given RC/CR network, it is determined that the system will yield maximum image-rejection at a single frequency point. This means that the IF should be a known frequency. If IF programmability is required, then ohmic resistors may be replaced by MOS transistors operating in triode, where the equivalent resistance is controlled by means of gate bias voltages.

Instead of passive phase shifters, active cells may be used. However, the latter may be problematic at high frequencies of operation. In Fig. 6.22, a 180^0 shifter is shown, based on the differential pair. Digital techniques may also be employed (Fig. 6.23), yet the operating frequency constraints apply in this case as well.

Returning to the implementation of the image-rejection mixer of Fig. 6.20, the measurement results are summarized in Table 6.12.

6.6.3   Image - Rejection Mixer with Automatic Gain Control

In the previous paragraph, it was mentioned that image rejection may be compromised if the IF frequency is displaced from a nominal value determined by the RC/CR network. To this extent a similar topology has been proposed [24-25], where an automatic gain control system (VCG) precedes the combiner, in order to equalize the signal amplitudes and provide optimal image rejection

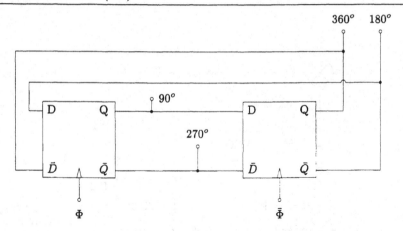

Figure 6.23: Digital phase shifter

Parameter	Measured Value
Application	DECT
Technology	BiCMOS 0.8μm
IF Frequency	111 MHz
Power Dissipation	60 mW
Image Rejection (IR)	14.1 dB
Conversion Gain	7.6 dB
NF	18 dB

Table 6.12: Performance of a BiCMOS IRM

across a broad IF frequency range. In Fig. 6.24, the system schematic is shown, comprising a VCO and a feedback circuit that implements the VCG. The multipliers employed in the system are conventional Gilbert cells. The RC/CR phase shifting networks are similar to those of Fig. 6.21(b). The combiner and voltage-controlled amplifier circuits are combined in one circuit depicted in Fig. 6.25.

The VCG control voltage is applied to the $M1 - M2$ transistor pair, which is biased by the $M5 - M7$ current mirror. The two current mirrors $M3 - M9$ and $M4 - M8$ control the transconductances of the differential pairs $Q1 - Q2$ and $Q3 - Q4$. In this way, the output voltage IF is controlled through the VCG mechanism.

The above system has been implemented in a $0.8\mu m$ BiCMOS process where the f_T for the npn transistor is 10 GHz. The use of the controlled gain amplifier results in a notable image rejection amount of 35 dB, for an IF frequency range from 150 to 250 MHz. The maximum rejection is at an IF

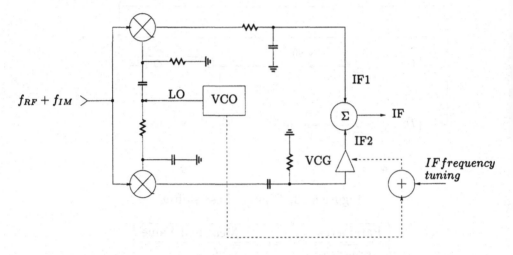

Figure 6.24: Image-rejection mixer with automatic gain control

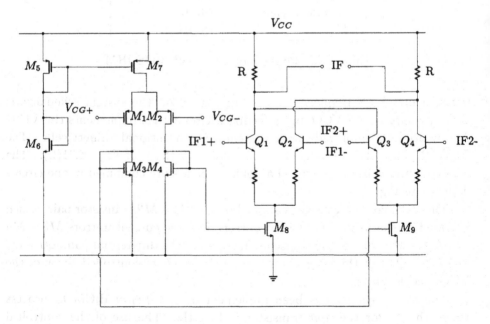

Figure 6.25: Combiner and voltage-controlled amplifier circuit

Parameter	Measured Value
Supply	3 V
RF Frequency	2 GHz
IF Frequency	200 MHz
Image Rejection	45 dB
Conversion Gain	8.5 dB
IIP3	−4.5 dBm
Current Dissipation	15 mA

Table 6.13: Performance of the IRM of Fig. 6.24

of 200 MHz. The system operates from a $3V$ supply and is aimed at 2 GHz applications. Table 6.13 summarizes the key measurement results.

The main cause for image rejection reduction is the mismatch between the signals, attributed to deviations from the nominal values of the circuit elements, that result in amplitude and phase discrepancies. Equation (6.19) provides an approximate relation between signal properties' mismatch and resulting image rejection :

$$\sqrt{(\text{amplitude error})^2 + (\text{phase error})^2} \simeq IRR \qquad (6.19)$$

where IRR is the image rejection ratio, i.e. the ratio of the unwanted over the useful signal power. For example, if a total error of 3% is considered, uniformly allocated between amplitude and phase, an amplitude error of $-33dB$ and a phase error of $\pm 2^0$ are attained. In this case, $IRR = 30dBc$.

6.7   I-Q MODULATORS

In digital modulation systems, the key cell is the I-Q modulator, the circuit that receives baseband signals (of zero and 90^0 phase) and produces the RF signal at the output. For RF signal generation, there are two main techniques: IF modulation and direct conversion. In IF modulation, the I-Q modulator outputs a signal in the 100-200 MHz range (for carriers up to 2 GHz) and subsequently the mixer upconverts this to the RF frequency (Fig. 6.26(a)). In direct conversion, the I-Q modulator outputs the RF signal directly (Fig. 6.26(b)).

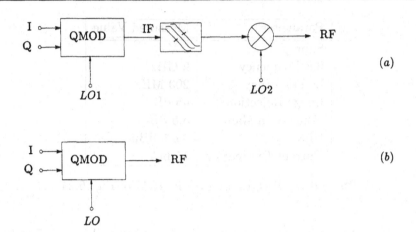

Figure 6.26: I-Q modulation techniques

6.7.1 Direct Conversion I-Q Modulator

The schematic of an I-Q modulator utilizing integrated phase shifters, is shown in Fig. 6.27 [26-27]. The circuit has been implemented in a $3V$ bipolar process. It comprises two doubly balanced mixers that receive the I and Q signals and are driven by balanced LO signals that are 90^0 out of phase between the two mixers.

The 90^0 phase difference is created by the phase shifter circuit $\Delta\phi$, which is controlled by voltage V_C. Two single-ended-to-balanced converters (S-B) provide a balanced LO signal to the mixers' inputs. Finally, a combiner circuit Σ at the system's output generates the RF signal.

In digital modulation systems, modulation accuracy is determined by the error vector in the phase domain. The error vector is defined as the ratio of magnitudes between the actual and ideal vectors. The main source of this error is the non-linear behaviour of doubly balanced mixers and the discrepancies (in amplitude and phase) between phase-shifted signals. In bipolar mixers, non- linearity is mainly accounted to third order products.

In Fig. 6.28, the proposed bipolar mixer topology [27] is shown. Due to the low supply voltage, a "current folding" technique has been employed, in order to reduce the number of series connected transistors.

The core of the circuit comprises two cross-connected balanced mixers ($Q1 - Q2$ and $Q3 - Q4$). The input voltage signal IN is converted to a current by a differential current mirror, comprising a linearized differential amplifier $Q5 - Q6$ and two current mirrors $Q7 - Q9$ and $Q10 - Q12$. The current sources $Q13$ and $Q14$ ensure that the sum of the two mixer currents

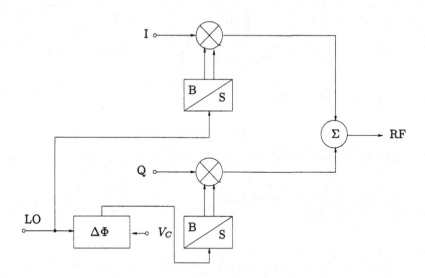

Figure 6.27: Direct conversion I-Q modulator

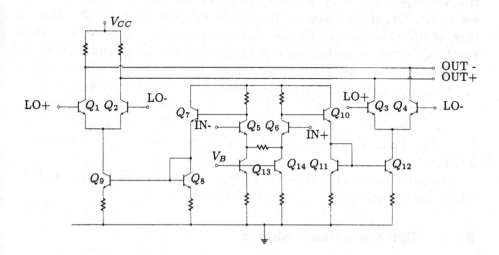

Figure 6.28: Doubly balanced mixer

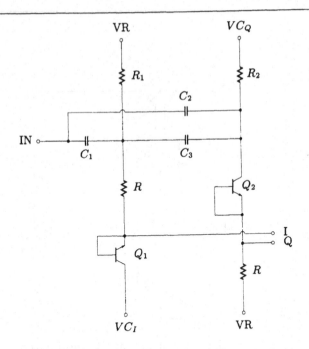

Figure 6.29: Phase shifter

remains constant, thereby yielding low distortion levels.

The use of current folding in this topology enables the use of only two transistors per branch. As a result, the circuit can operate from a supply as low as $2V$. The utilized phase shifter circuit is depicted in Fig. 6.29. This is a tunable RC/CR network. The problem with plain RC/CR networks with no tunability is that, if the bandpass and highpass filters have identical R and C elements, the phase difference between I and Q at the output will be (90^0) at any frequency, the signal amplitudes however will change with frequency.

This lead to the topology of Fig. 6.29, where transistors $Q1$ and $Q2$ are used as variable capacitors (B-C junction), controlled by voltages VC_I and VC_Q. This permits an accurate phase shifter operation in a range from 0.8 to 2 GHz.

The circuit has been designed in a bipolar technology where $f_T = 18$ GHz. The measurement results are given in Table 6.14.

6.7.2   High Gain Phase Shifter

The basic configuration of an active high-gain phase shifter appears in Fig. 6.30 [28-29]. The common-base transistor $Q1$ provides the necessary 50Ω matching through its biasing, since the input resistance R_{in} is approximately

Parameter	Measured Value
Operating Frequency	$0.8 \div 2$ GHz
Image Rejection	< -35 dBc
Second and Third Order Products	< -40 dBc
LO Leakage	< -40 dBc
Error vector	4.9 % rms
Supply	2 V
Power Dissipation	68 mW

Table 6.14: Performance of the I-Q modulator of Fig. 6.27

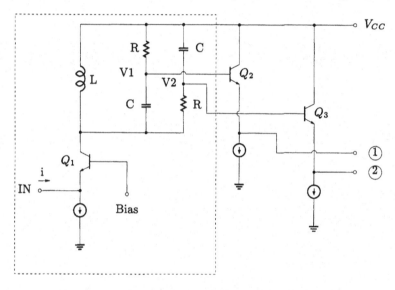

Figure 6.30: Active phase shifter

equal to the $\frac{1}{g_m}$ of $Q1$. Transistor $Q1$ also drives the RC/CR network with current. The collector current of $Q1$ is almost equal to the input current, due to the common base configuration. Therefore the output voltage of the phase shifter v_1 is approx. equal to $i\frac{R}{2}$ (provided that inductance L is sufficiently large). Respectively, the output voltage of a regular phase shifter is $iR_{in}/\sqrt{2}$, independent of the R and C values. Thus, an appropriate R value, $(R > \sqrt{2} \cdot R_{in})$, will yield a higher output voltage for this phase shifter.

At this point the presentation of silicon integrated mixer circuits is concluded. In Chapter 8, integrated RF filters will be described and the reader will be able to compare the performance of mixers combined with image- rejecting filters, against that of integrated image-rejection mixers (IRM).

Parameter	Measured Value
Operating frequency	0.8 – 2 GHz
Image Rejection	< –35 dBc
Second and Third Order Products	< –40 dBc
LO leakage	< –40 dBc
Error vector	4.0 % rms
Supply	5 V
Power Dissipation	65 mW

Table 6.15. Performance of the I-Q modulator of Fig. 6.27.

Figure 6.26. Active phase shifter.

Bibliography

[1] D.O. Pederson and K. Mayaram, *Analog Integrated Circuits for Communication*, Kluwer Academic Publishers, 1991.

[2] K.L. Fong and R.G. Meyer, "A 2.4 GHz Monolithic Mixer for Wireless LAN Applications," *IEEE 1997 CICC, Paper 9.4.1.*, pp. 185–188.

[3] R.G. Meyer and W.D. Mack, "A 1-GHz BICMOS RF Front-End I.C.," *IEEE J. of Solid-State Circuits*, vol. 29, no. 3, pp. 350–355, March 1994.

[4] S.A. Maas, *Microwave Mixers*, Artech House, 1993.

[5] J.R. Long and M.A. Copeland, "A 1.9 GHz Low-Voltage Silicon Bipolar Receiver Front-End for Wireless Personal Communication Systems," *IEEE J. of Solid-State Circuits*, vol. 30, no. 12, pp. 1438–1448, Dec. 1995.

[6] T. Yamaji and H. Tanimoto, "A 2 GHz Balanced Harmonic Mixer for Direct-Conversion Receivers," *IEEE 1997 CICC Paper 9.6.1.*, pp. 193–196.

[7] K.L. Fong, C.D. Hull, and R.G. Meyer, "A Class AB Monolithic Mixer for 900 MHz Applications," *IEEE J. of Solid-State Circuits*, vol. 32, no. 8, pp. 1166–1172, Aug. 1997.

[8] B. Razavi, "A 1.5 V 900 MHz Downconversion Mixer," *IEEE ISSCC 96 Paper TP 3.1*, pp. 48–49.

[9] B. Gilbert, "Mixer Fundamentals and Active Mixer Design," *RF IC Design for Wireless Communication Systems*, EPFL, Switzerland, 1995.

[10] G.C. Dowe, J.M. Mourant, and A.P. Brokaw, "A 2.7-V DECT RF Transceiver with Integrated VCO," *IEEE ISSCC 97, Paper SA 18.5*, pp. 308–309.

[11] B. Gilbert, "The MICROMIXER: A Highly Linear Variant of the Gilbert Mixer Using a Bisymmetric Class - AB Input Stage," *IEEE J. of Solid-State Circuits*, vol. 32, no. 9, pp. 1412-1423, Sept. 1997.

[12] J.C. Rudell, J.-J.Ou, T.B. Cho, G. Chien, F. Brianti, J.A. Weldon, and P.R. Gray, "A 1.9 GHz Video-Band IF Double Conversion CMOS Integrated Receiver for Cordless Telephone Applications," *IEEE ISSCC 97, Paper SA 18.3*, pp. 304-305.

[13] A. Rofougaran, J. Y.-C. Chang, M. Rofougaran, and A.A. Abidi, "A 1 GHz CMOS RF Front-End IC for a Direct-Conversion Wireless Receiver," *IEEE J. of Solid-State Circuits*, vol. 31, no. 7, pp. 880-889, July 1996.

[14] A.R. Shahani, D.K. Shaeffer, and T.H. Lee, "A 12 mW Wide Dynamic Range CMOS Front-End for a Portable GPS Receiver," *IEEE ISSCC 97, Paper SP. 22.3*, pp. 368-369.

[15] J. Crols and M. Steyaert, "A Full CMOS 1.5 GHz Highly Linear Broadband Downconversion Mixer," *ESSCIRC 94*, pp. 248-251.

[16] J. Crols and M. Steyaert, "A fully Integrated 900 MHz CMOS Double Quadrature Downconverter," *IEEE ISSCC 95, Paper TA 8.1*, pp. 136-137.

[17] P.Y. Chan, A. Rofougaran, K.A. Ahmed, and A.A. Abidi, "A Highly Linear 1-GHz CMOS Downconversion Mixer," *ESSCIRC 93*, pp. 210-213.

[18] B. S. Song, "CMOS RF Circuits for Data Communications Applications," *IEEE J. of Solid-State Circuits*, vol. SC-21, no. 2, pp. 310-317, April 1986.

[19] A. N. Karanicolas, "A 2.7 V 900 MHz CMOS LNA and Mixer," *IEEE ISSCC 96, Paper TP 3.2*, pp. 50-51.

[20] A. N. Karanicolas, "A 2.7 -V 900-MHz CMOS LNA and Mixer," *IEEE J. of Solid - State Circuits*, vol. 31, no. 12, pp. 1939-1944, Dec. 1996.

[21] R.G. Meyer, W.D. Mack, and J. Hageraats, "A 2.5 GHz BICMOS Transceiver for Wireless LAN," *IEEE ISSCC 97, Paper SA 18.6*, pp. 310-311.

[22] J.R. Long, M.A. Copeland, S.J. Kovacic, D.S. Malhi, and D.L. Harame, "RF Analog and Digital Circuits in SiGe Technology," *IEEE ISSCC 96, Paper TP 5.3*, pp. 82-83.

[23] M.D. McDonald, "A 2.5 GHz BICMOS Image-Reject Front-End," *ISSCC 93, Paper TP 9.4*, pp. 144–145.

[24] D. Pache, J.M. Fournier, G. Billiot, and P. Senn, "An Improved 3V 2GHz BICMOS Image Reject Mixer IC," *IEEE CICC 95, Paper 6.3.1*, pp. 95–98.

[25] D. Pache, J.M. Fournier, G. Billiot, and P. Senn, "An Improved 3V 2GHz Image Reject Mixer and a VCO-Prescaler Fully Integrated in a BICMOS Process," *IEEE CICC 1996.*

[26] T. Tsukahara, M. Ishihawa, and M. Muraguchi, "A 2V 2GHz Si-Bipolar Direct-Conversion Quadrature Modulator," *IEEE ISSCC 94, Paper WP 2.6*, pp. 40–41.

[27] T. Tsukahara, M. Ishihawa, and M. Muraguchi, "A 2-V 2-GHz Si-Bipolar Direct-Conversion Quadrature Modulator," *IEEE J. of Solid - State Circuits*, vol. 31, no. 2, pp. 263–267, Feb. 1996.

[28] M. Ishibe et al., "High-Speed CMOS I/O Buffer Circuits," *IEEE J. of Solid - State Circuits*, vol. 27, no. 4, pp. 671–673, Apr. 1992.

[29] S. Otaha, T. Yamaji, R. Fujimoto, C. Takahashi, and H. Tanimoto, "A Low Local Input 1.9 GHz Si-Bipolar Quadrature Modulator with No Adjustment," *IEEE J. of Solid - State Circuits*, vol. 31, no. 1, pp. 30–37, Jan. 1996.

CHAPTER
7

OSCILLATORS

7.1 INTRODUCTION

The frequency synthesizer is one of the most fundamental cells in a telecommunications transceiver. Its role is to produce the necessary periodic signals for frequency up-conversion in the transmitter and down-conversion in the receiver. The location of the frequency synthesizer in a typical transceiver is depicted in the schematic of Fig. 7.1.

The local oscillator (LO) is an integral part of the frequency synthesizer, characterized by very stringent design requirements. The produced frequency should remain constant and precisely fixed, with an accuracy ranging from 0.1 ppm for GSM to 25 ppm for DECT applications. Moreover, the LO frequency should be adjustable in small accurate steps, to enable switching between channels. These channels are spaced at 200 kHz for GSM and 1.728 MHz for DECT, while in other cases channel spacing may be even smaller (e.g. 30 kHz for IS-54). To achieve the necessary LO frequency accuracy, a crystal oscillator is usually employed as a high precision frequency reference. However, environmental conditions and aging may shift this reference. If the system specifications call for extremely high accuracy, an automatic frequency control (AFC) system should be employed for error correction.

The increased accuracy requirements associated with the frequency synthesizer necessitate the utilization of phase-locked loop (PLL) systems for

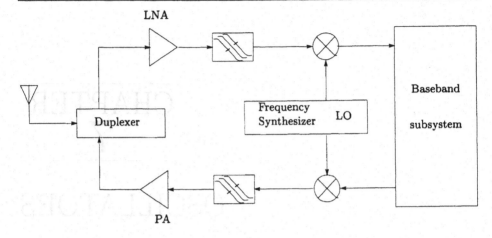

Figure 7.1: Transceiver schematic

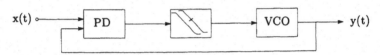

Figure 7.2: PLL schematic

frequency synthesis in actual telecommunications systems.

The basic topology of a PLL is shown in Fig. 7.2. This is a feedback system that controls the phase difference between the generated and the reference signal. The PLL system comprises a phase detector (PD), a lowpass filter and a voltage-controlled oscillator (VCO). When the input and output frequencies are equal, the phase difference $\Delta\phi$ remains constant with time. Then the PLL is in the "locked" state. In order for the PLL to reach this state, the phase detector produces an output voltage with a dc component proportional to $\Delta\phi$. The subsequent lowpass filter stifles high frequencies and produces a dc voltage which is used in controlling the VCO. In this fashion, when frequencies of signals x(t) and y(t) coincide, the VCO control voltage remains constant and the loop is locked.

There are various well-known PLL architectures [1,2] and the most fundamental will be presented soon after. However, the emphasis of this chapter is placed on integrated VCO implementations.

7.2 FREQUENCY SYNTHESIZERS OVERVIEW

a) Integer-N Architecture

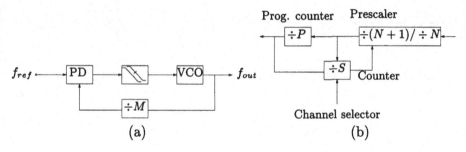

Figure 7.3: Integer-N PLL

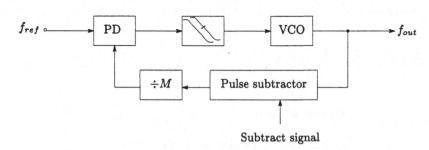

Figure 7.4: Fractional-N PLL

A PLL with a programmable divider in the feedback path, as in Fig. 7.3(a), constitutes a complete frequency synthesizer circuit. In Fig. 7.3(b), the schematic of the divider circuit is shown, comprising a dual modulo prescaler and two counters. The divider operates in such a way that a pulse appears at its output every $(N+1)S + (P-S)N = NP + S$ VCO pulses.

The simplicity of this particular architecture makes it a popular choice in RF systems design. Furthermore, the design of the PLL in this case can be broken down to three separate integrated circuits: a VCO on one chip, a prescaler and two counters on a second, a phase detector and a lowpass filter on a third.

b) Fractional-N Architecture

In fractional-N systems the output frequency can be tuned at a fraction of the reference frequency, allowing for f_{ref} values much larger than the channel spacing in the system. The PLL schematic is shown in Fig. 7.4.

The pulse subtractor in the circuit of Fig. 7.4 removes one input signal pulse when the subtraction signal is active. In the locked state the two frequencies are equal and therefore f_{out} equals $M f_{ref} + \frac{1}{T}$ where T is the period during which the subtraction signal is applied. For example, if for a certain time the VCO output is divided by M and for some other time it is divided

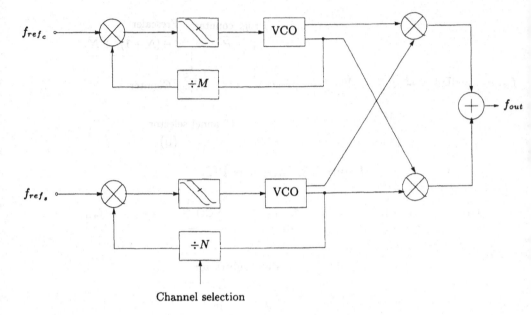

Figure 7.5: Double loop synthesizer

by $M+1$, then the average of the division factor lies between M and $M+1$.

Therefore, in this case the combination of pulse subtractor and divider by M becomes a dual modulo prescaler.

c) Double Loop Architectures

Sometimes it is desirable to synthesize frequencies that can be tuned in very small steps. This can be achieved by a technique that involves adding a small tunable frequency to a high constant frequency. The technique employs two PLLs, as depicted in Fig. 7.5. The first PLL produces a carrier frequency and the second produces the small frequency steps. A mixer is used to add the two frequency components.

Many variations of the double loop topology of Fig. 7.5 have been reported.

d) Direct Digital Synthesis (DDS)

The direct digital synthesis technique differs from conventional PLL techniques since it is based on purely digital signal generation. In this case, a D/A converter and filter compose the analog waveform. The DDS technique is presented in Fig. 7.6. The register produces digital step waveforms that are mapped by the ROM to a digital sinusoid. The DAC and filter circuits consequently convert the sinusoid to analog.

The DDS technique exhibits many advantages in comparison to PLL architectures, such as lower phase noise, better resolution and faster channel

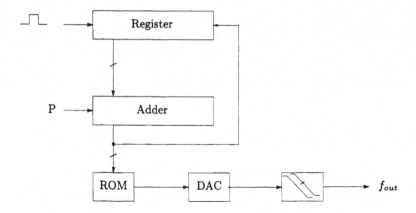

Figure 7.6: DDS architecture

switching. A serious drawback - with respect to RF applications - is that it requires a high frequency of operation. The clock used in the register should be three to four times faster than the produced sinusoid. For example, in order to produce a GSM carrier frequency the digital circuits would operate at 3 GHz! This is prohibitive even for the most advanced VLSI technologies (1997) - especially considering the power consumption it would lead to. The design of the DAC circuit would also pose a serious challenge.

7.3 PHASE NOISE

Apart from frequency stability, another critical parameter of oscillators and frequency synthesizers is their phase noise. If the LO signal exhibits phase noise, then this noise will pass through the mixer and will be up-converted to the RF band (at the transmitter) or down-converted to the baseband (at the receiver). Phase noise can be defined either in the time domain or in the frequency domain [3]. In the time domain it appears as clock jitter that can hinder the clock recovery process. The output voltage of the oscillator is given by:

$$V_{osc} = V_A \cdot \cos[\omega t + \phi(t)] \tag{7.1}$$

where $\phi(t)$ is random noise. In Fig. 7.7 the respective waveform is displayed. Accordingly, in the frequency domain the oscillator produces an output spectrum such as the one shown in Fig. 7.8. As it can be seen, besides a frequency component at ω_0 a continuous spectrum around it is also produced, containing

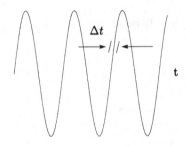

Figure 7.7: Phase noise in the time domain.

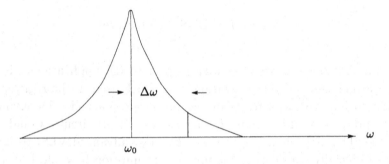

Figure 7.8: Phase noise in the frequency domain.

a certain power at a distance $\Delta\omega$ from ω_0. The power of these sidebands hampers frequency synthesis, especially if the channels in the system are spaced closely together (e.g. GSM). In telecommunications systems design, phase noise is usually considered in the frequency domain. In the receiver case, the problem is illustrated in Fig. 7.9: If the oscillator's output has the profile of Fig. 7.8, at the mixer's RF input two signals will be applied, the desired signal at ω_1 and a strong unwanted signal at ω_2, as shown in Fig. 7.9(a). The result of mixing these signals is shown in Fig. 7.9(b). The unwanted signal has been down-converted to the baseband and has clearly distorted the useful information.

Similar problems are created at the transmitter side, where the phase noise of the LO results in the transmission of unwanted sidebands instead of a "clean" signal. These unclean signals may cause problems to neighboring receiver circuits.

The units for quantifying phase noise can be directly derived from its representation in the frequency domain. Therefore, phase noise is expressed

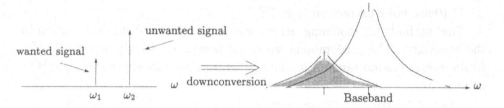

Figure 7.9: Effect of oscillator phase noise on the receiver system.

in dBc/Hz, according to the following equation:

$$\text{Phase Noise} = \mathcal{L}\{\Delta\omega\}$$
$$= 10\log\frac{\text{Noise power in 1 Hz bandwidth at frequency } \omega_0 + \Delta\omega}{\text{Carrier power}} \quad (7.2)$$

From equation (7.2) it is obvious that apart from the absolute value of phase noise, the offset from the carrier frequency at which it is measured should also be mentioned. This depends on the exact application, as it was mentioned in Paragraph 7.1. A typical quoted value for telecommunications applications is -80 dBc/Hz @ 10 kHz, or a signal level difference of 80 dB between the carrier frequency and the 10 kHz offset frequency.

The main issue with phase noise prediction in electronic simulation, arises from the non-linear nature of the oscillator: Unfortunately, conventional ac noise simulation cannot be applied here, and a reliable method that would yield simulation results close to measured data has yet to be established. This means that several iterations may be needed, creating a time and cost overhead in oscillator design. To this extent, a rigorous research effort is underway, with no definitive results up to now (1997). Also, some oscillator types are not adequately covered by reported techniques. For example, the technique employed for phase noise simulation in SpectreRF, shows considerable deviations from measured results - especially in the case of ring oscillators. Recently, a new phase noise calculation scheme has been reported [4], exhibiting a noteworthy simulation accuracy (1 dB discrepancy for a 1.8 GHz VCO). This scheme simulates noise in the time domain, in the following steps:

i) Assignment of all noise sources in the circuit (if integrated inductors are employed, they should be carefully taken into account since there is no systematic modeling of their behavior).

ii) Modeling of all noise sources and assignment of frequencies of interest.

iii) Noise simulation in the time domain, including the noise sources.

iv) FFT of the results of time domain simulation.

v) Phase noise extraction from FFT.

The methods for modeling noise sources have received much attention in the literature. The predominant views (although not widely accepted) refer to the use of pseudo-random number generators or pure sinusoidal signals as noise sources.

Other reported techniques propose the use of a simplified version of the oscillator circuit and the subsequent application of analytical calculations to it [5-9]. A recent technique employs a Monte Carlo algorithm for simulating noise in the time domain [10]. The method is generic and can be applied to all kinds of oscillators.

7.4   OSCILLATOR TYPES

Many types of oscillators have been inherited from discrete component circuit design for telecommunications systems. In this paragraph, some key architectures will be portrayed, while the existing integrated implementations will be presented next. The following oscillator categories can be identified:

- Tuned-LC (harmonic) oscillators.
- Crystal oscillators.
- OTA-C oscillators.
- Ring oscillators.
- Relaxation oscillators.

The most common oscillators in integrated form are of the tuned-LC variety. Usually, the LC tank is placed off-chip. Recently, there has been some effort to integrate the tuned LC tank using capacitors, varactors and integrated inductors. Facing the obstacle of passive integrated inductor performance and modeling, designers have often resorted to the use of "active inductors", implemented by OTAs and capacitors as depicted in Fig. 7.10. Obviously such a solution entails several problems, mainly increased noise and power dissipation - especially at high frequencies.

7.5   BIPOLAR OSCILLATORS

7.5.1   Varactor-Tuned Bipolar Oscillator

The schematic of an oscillator in a bipolar technology is shown in Fig. 7.11 [11]. A varactor is employed for center frequency tuning and an inductor is used externally to minimize losses. The external component has an adequately

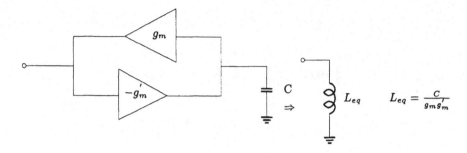

Figure 7.10: Active inductors' principle (for use in OTA-C oscillators)

high quality factor, so that the total quality factor of the LC tank can be at least equal to that of the varactor (higher than 10).

The oscillator employs an amplifier comprising two transistors $(Q_1 - Q_2)$ in an ECP configuration, that exhibits optimum performance in the area of 0.1 to 1 GHz. The ECP configuration offers the advantage of isolating the LC tank from the oscillator output, as well as a high input resistance. Also, the total harmonic distortion is lower than in the single-transistor case. Positive feedback is provided through capacitors C_1 and C_2.

This oscillator is part of an integrated PLL that operates at a nominal frequency of 350 MHz. An external inductor made of silver was used, having an inductance of 5 nH and a quality factor of 10.

7.5.2 Integrated Bipolar Oscillator at 250 MHz

The schematic of an oscillator with double loop feedback for frequency stabilization is shown in Fig. 7.12 [12].

The elements F_i in the circuit of Fig. 7.12 are frequency-to-voltage converters: when the circuit settles, the output voltage of F_i becomes equal to the reference voltage (within a finite error margin). In the double control loop, there is a second reference loop that tracks the deviations of F_i elements under temperature and other changes. Any change in F_2 is amplified by A_2 and its output V_B is fed back to F_i, so that V_C remains constant and equal to reference voltage V_{ref}. The input at F_2 is a constant frequency f_R. Feeding the voltage V_B to F_1, frequency is converted to a stabilized voltage, independent of temperature or other conditions. Temperature stabilization is accomplished by the second loop, which is separate from the first, so that it does not affect signal flow.

Reference frequency f_R is produced by a crystal oscillator, while the VCO is an emitter-coupled oscillator (ECO). The circuit diagram of the oscillator

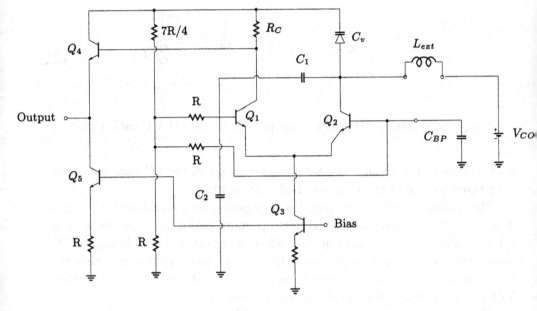

Figure 7.11: Varactor-tuned oscillator

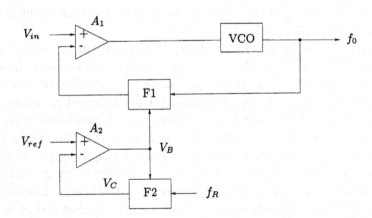

Figure 7.12: Double loop feedback oscillator schematic

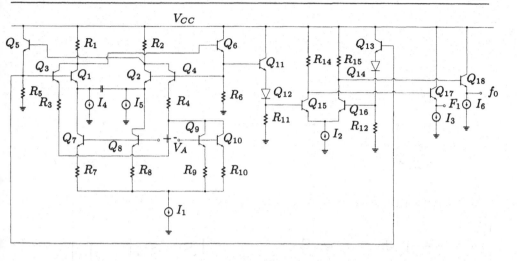

Figure 7.13: Emitter-coupled VCO

Parameter	Measured Value
Operating Frequency	250 MHz
TC / Frequency	-65 ppm/$^\circ C$
Supply Voltage	5V
Power consumption	220 mW

Table 7.1: Measurement results of the system of Fig. 7.12

is shown in Fig. 7.13. Resistors R_3 and R_4 are used to induce oscillations and avoid locking at an unwanted state.

The current sources I_4 and I_5 are employed to avoid oscillation termination if V_A drops below a certain value. Emitter followers Q_5 and Q_6 buffer the two square wave outputs from the oscillator to the differential amplifier Q_{15}-Q_{16}, for further signal conditioning. Experimental results of the double loop feedback arrangement are summarized in Table 7.1.

7.5.3   Tuned-LC Oscillator at 1.8 GHz

In Fig. 7.14, the schematic of a tuned-LC VCO is shown [13]. This specific topology utilizes only on-chip elements and is one of the first reported designs with integrated inductors on silicon.

The circuit comprises two LC tanks with different resonance frequencies. The feedback loop is implemented in a conventional Colpitts topology. A Gilbert cell (Q_1-Q_4) is used for tuning the oscillation frequency of the VCO through voltages V_C^+ and V_C^-. Thus, oscillation frequencies between the self-

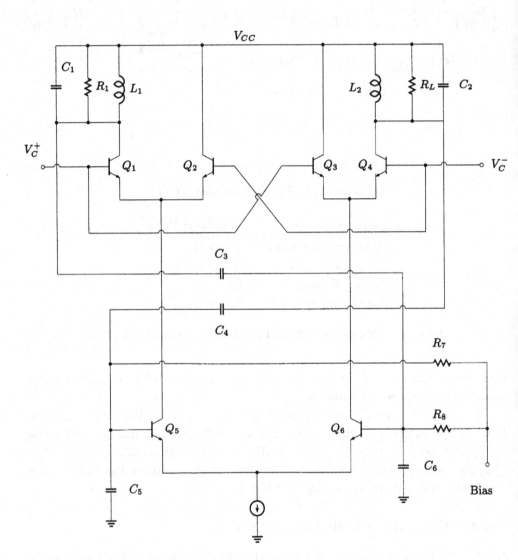

Figure 7.14: Tuned-LC VCO

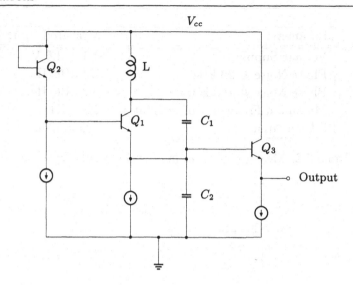

Figure 7.15: Colpitts-type integrated oscillator

resonance frequencies of the two LC tanks can be attained.

For the implementation of this circuit, integrated inductors with values of $L_1 = 6.5$ nH and $L_2 = 3.7$ nH were used. Capacitor C_1 had a low value (0.2pF), while for C_2 the parasitics of the inductors were exploited. Also, resistors R_1 and R_2 correspond to the inductors' ohmic losses. The VCO operates at a nominal frequency of 1.8 GHz and can be tuned in a 200 MHz range. The quiescent power dissipation is 70 mW, drawn from a 5V supply. The phase noise at a 20 kHz offset from the carrier ($f_0 = 1.68$ GHz) is relatively high: - 67 dBc/Hz. The circuit was implemented in a BiCMOS technology where the f_τ of the npn transistor was 10 GHz.

7.5.4 Integrated Resonant Circuit Oscillator at 2.4 GHz

The simplified schematic of a Colpitts oscillator that employs an integrated inductor, is shown in Fig. 15 [14].

In this particular implementation, the resonant circuit comprises an integrated inductor L and a capacitive transformer C_1/C_2 with a ratio of 2:1. The emitter follower Q_3 buffers the produced waveform from the middle point of the transformer, so that the voltage at the terminals of the resonant circuit is not affected.

To achieve a high quality factor for the integrated inductor, a four-turn spiral was used, laid out on three metal layers ($M2 \| M3 \| M4$) that were connected together in order to minimize ohmic losses. This scheme yielded an

Parameter	Measured value
Voltage Supply	2.6 ÷ 3.6 V
Phase Noise @ 20 kHz	-78 dBc/Hz
Phase Noise @ 100 kHz	-92 dBc/Hz
Maximum Frequency of Oscillation	2.36 GHz
Silicon Area	$500 \times 500 \mu m^2$

Table 7.2: Measurement results for the oscillator of Fig. 7.15

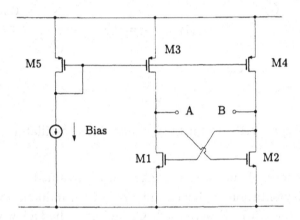

Figure 7.16: CMOS VCO - active circuit

inductor with an inductance value of 1.9 nH and Q = 11 at a frequency of 2.4 GHz. The circuit was implemented in a BiCMOS process where the f_τ of the bipolar element was 12 GHz. The experimental measurement results are summarized in Table 7.2.

7.6   CMOS OSCILLATORS

7.6.1   Low Noise Tuned-LC Oscillator

The active circuit of an MOS VCO in common source configuration is shown in Fig. 7.16 [15]. Transistors M1 and M2 are cross-connected in order to provide positive feedback. The circuit operates from a 3V supply and produces an adequately high-amplitude oscillation, so that the phase noise can be kept at low levels. High frequency operation is enhanced by the fact that the parasitic capacitances C_{gs} of M1-M2 appear in parallel to the capacitors of the LC tank, which is connected between points A and B as shown in Fig. 7.16. The LC network used in this case is displayed in Fig. 7.17.

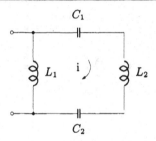

Figure 7.17: LC network for the VCO of Fig. 7.16

Parameter	Measured Value
Oscillation Frequency	1.76 GHz
Phase Noise @ 10 kHz	-85 dBc/Hz
Phase Noise @ 200 kHz	-115 dBc/Hz
Voltage Supply	3V
Current Consumption	8 mA
Tuning Range	4.5 %

Table 7.3: Performance of the VCO of Fig. 7.16

This modified LC network exhibits some interesting properties: The current i flowing in the loop of Fig. 7.17 inflicts a voltage drop on inductors L_1 and L_2 and a voltage increase on C_1 and C_2, due to the negative reactance of the capacitors. If the same current was flowing through a simple LC branch, with L=L_1+L_2 and $C = (C_1^{-1} + C_2^{-1})^{-1}$, the voltage drop across it would be twice as large. Therefore, since the noise level is the same in both networks, it is inferred that the modified LC tank leads to a 6 dB decrease in phase noise.

For this specific VCO implementation, bonding wires were used to yield low ohmic loss inductors and thus achieve a phase noise as low as possible. In order to obtain the large required inductance values, long bondwires that ran across the packaging of the chip were utilized.

A combination of MIM capacitors (for low parasitic resistance) and p+/n - well junction capacitors as varactors (for frequency tuning) was used. The circuit was fabricated in a CMOS $0.7\mu m$ technology. Table 7.3 summarizes its performance.

7.6.2 CMOS Oscillator at 4 GHz

The simplified schematic diagram of a CMOS VCO that uses integrated inductors, capacitors and varactors [16] is shown in Fig. 7.18. All MOS transistors

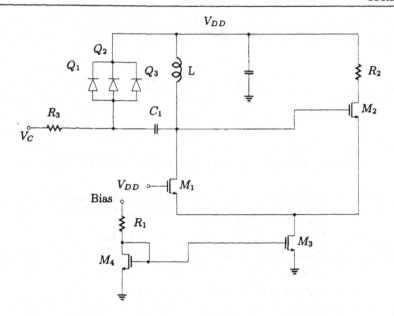

Figure 7.18: Schematic of a VCO at 4 GHz

have been designed with multiple gates, so as to minimize the parasitic ohmic resistance of the gate, as well as the diffusion capacitances. The required positive feedback is implemented by the connection of the gate of M2 to the drain of M1. This is also the node where the tuned-LC circuit is connected. Transistors M3 and M4 form the biasing circuit through a current mirror of ratio 1:1.

The integrated inductor is a simple four-turn square spiral, of a 2.4 nH nominal value. The capacitor C_1 is a MIM capacitor of a 5 pF nominal value. The varactors are implemented by the base-collector junction of the bipolar transistors Q1-Q3. The varactor of Q_1-Q_3 exhibits a capacitance of 0.5 pF, with a quality factor of 5. The negative resistance realized by the active circuit has a value of -57 Ω at 4 GHz, for a biasing current of 4 mA. The whole circuit was implemented in a $0.5\mu m$ BiCMOS process, with five layers of metallization. The nMOS transistors used had an f_T of 12 GHz. Experimental measurements of the integrated circuit have revealed the results found in Table 7.4. The relatively high phase noise is partly due to the low quality factor of the passive elements at 4 GHz.

Parameter	Measured Value
Oscillation Frequency	4 GHz
Tuning Range	9 %
Supply Voltage	3V
Current Consumption	4 mA
Phase Noise @ 100 kHz	-85 dBc/Hz
Phase Noise @ 1 MHz	-106 dBc/Hz

Table 7.4: Performance of the VCO of Fig. 7.18

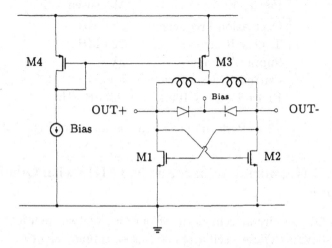

Figure 7.19: VCO at 1.8 GHz

7.6.3   Low Noise CMOS Oscillator at 1.8 GHz

In [17], a CMOS VCO that uses integrated inductors on two levels of metallization is presented. The circuit diagram of the oscillator is shown in Fig. 7.19. This is a well-known topology of a tuned-LC circuit, where two transistors M1-M2 connected in positive feedback provide a negative resistance. The inductors have a value of 3.2 nH, meaning that for oscillations at 1.8 GHz the total capacitance per node should be 2.4 pF. The parasitic ohmic resistance of the tuned-LC circuit is about 15 Ω and the required negative conductance from the active part of the oscillator is 3 mS. The capacitance of the LC network is provided by the parasitics of the integrated inductors, the parasitic capacitance of the MOS and by the variable capacitance of the p+/n-well junction diode that is implemented in the integrated circuit.

The circuit was fabricated in a 0.7μm digital CMOS process. The measurement results are given in Table 7.5.

Parameter	Measured Value
Oscillation Frequency	1.8 GHz
Tuning Range	14 %
Supply Voltage	1.5V
Power dissipation	6 mW
Phase Noise @ 600 kHz	-116 dBc/Hz

Table 7.5: Performance of the oscillator of Fig. 7.19

Parameter	Measured Value
Oscillation Frequency	900 MHz
Tuning Range	120 MHz
Supply Voltage	3V
Current Consumption	5 mA
Phase Noise @ 100 kHz	-85 dBc/Hz

Table 7.6: Performance of the oscillator of Fig. 7.20

7.6.4 LC-Network Oscillator at 900 MHz with Quadruple Output

In Fig. 7.20 the circuit schematic of one out of two identical CMOS oscillators is shown. These oscillators are cross-connected in order to produce four-quadrant LO signals [18]. The topology is basically the same with that of the previous paragraph.

The frequency of oscillations is controlled by voltage V_C. To increase the quality factor of the integrated inductors, the substrate underneath them has been removed. This reduces the parasitic effects. The circuit was fabricated in a $1\mu m$ CMOS technology and the experimental results are summarized in Table 7.6.

A similar circuit to the one above is presented in [19]. In that case an inductor laid out on two layers of metallization has been used, in order to save silicon area. The circuit was implemented in a $0.6\mu m$ CMOS process and its performance is given in Table 7.7.

7.6.5 Inductorless CMOS Oscillators

An interesting effort was presented in [20], where three topologies of VCOs without inductors are proposed, exhibiting fairly good noise performance. The first one is a modified ring oscillator, shown in Fig. 7.21. The delay circuit

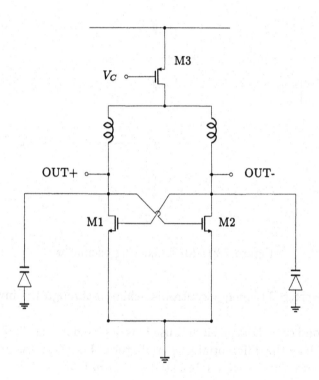

Figure 7.20: VCO at 900 MHz

Parameter	Measured Value
Oscillation Frequency	1.8 GHz
Tuning Range	120 MHz
Supply Voltage	3.3 V
Current Consumption	2.3 mA
Phase Noise @ 500 kHz	-100 dBc/Hz

Table 7.7: Performance of an oscillator at 1.8 GHz

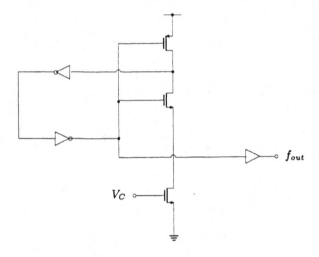

Figure 7.21: Modified ring oscillator

is a simple inverter. Frequency control is achieved through the inverter's bias current.

In the second case, D-type latch is used, as depicted in Fig. 7.22. Finally, in the third topology the differential ring oscillator of Fig. 7.23 was implemented.

All the above circuits were fabricated in $1.2\mu m$ CMOS. Table 7.8 summarizes the experimental results.

A similar design at 2 GHz is found in [21]. This one was implemented in a BiCMOS process where f_τ=20 GHz for the npn transistor.

7.7   SiGe OSCILLATORS

An impressive VCO design at 11 GHz in a SiGe technology, appears in [22]. Using heterojunction bipolar transistors (HBT), high-Q MIM capacitors, varactors and integrated inductors, an oscillator with very good noise performance was designed. Its simplified schematic is shown in Fig. 7.24. This is a Clapp

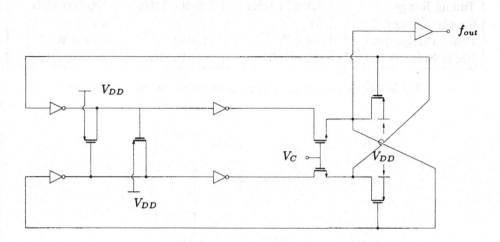

Figure 7.22: D-type latch VCO

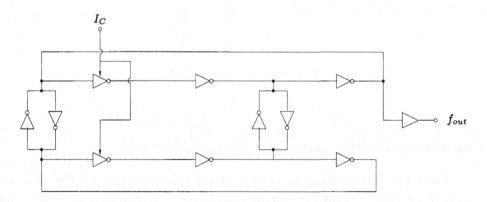

Figure 7.23: Differential ring oscillator

Parameter	Measured Value (Fig. 7.21)	Measured Value (Fig. 7.22)	Measured Value (Fig. 7.23)
Tuning Range	320-926 MHz	225-600 MHz	100-500 MHz
Supply Voltage	5V	5V	5V
Power Dissipation	9.4 mW	14 mW	20.2 mW
Phase Noise @ 100 kHz	-80 dBc/Hz	-87 dBc/Hz	-90 dBc/Hz

Table 7.8: Measurement results of inductorless oscillators

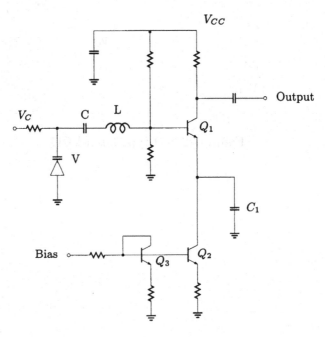

Figure 7.24: SiGe oscillator

oscillator modification where all the elements are on-chip.

The negative resistance is realized by the transistor Q_1 and the feedback capacitor C_1, along with the current source Q_2. The voltage-controlled circuit is implemented by the inductor L and the varactor V. The inductor is laid out on the topmost metallization layer (the third) and has a nominal value of 1.2 nH. The varactor is a base-collector bipolar junction and its capacitance is tuned through V_C. The experimental results for this circuit are summarized in Table 7.9.

Parameter	Measured Value
Oscillation Frequency	11 GHz
Tuning Range	5 %
Supply Voltage	3V
Current Consumption	8 mA
Phase Noise @ 100 kHz	-78 ÷ -87 dBc/Hz
Phase Noise @ 1 MHz	-98 ÷ -106 dBc/Hz

Table 7.9: Performance of the oscillator of Fig. 7.24

Parameter	Measured Value
Oscillation Frequency	714 MHz
Tuning Range	14 MHz
Supply Voltage	3.3 V
Current Consumption	15 mA
Phase Noise @ 100 kHz	-107 dBc/Hz

Table 7.10: Performance of the oscillator of Fig. 7.25

7.8 OSCILLATORS USING MICRO MECHANICAL CAPACITORS

In [23], an integrated Colpitts oscillator utilizing a micromechanical variable capacitor in a typical silicon technology is presented. The capacitor comprises a thin aluminum plate that is suspended with springs at its four corners over another aluminum plate. The use of aluminum yields a low ohmic resistance an thus a higher Q for the element. For a distance of $1.5\mu m$ between the plates, a nominal capacitance value of 200 fF is attained. When a bias voltage is applied at the capacitor's terminals, an electrostatic force pushes the upper plate downwards, thereby decreasing the distance between the plates. The maximum distance variance is approximately 1/3 of its initial value, resulting in a capacitance increase of about 50%.

The circuit diagram of the Colpitts oscillator is shown in Fig. 7.25. The first version of the proposed circuit implements a CMOS IC for the active elements and a separate CMOS IC for the micromechanical capacitors. The inductor is a discrete component. However, there seems to be no serious obstacle for the complete integration of this circuit on a common silicon substrate. The performance of the oscillator is summarized in Table 7.10.

Concluding the presentation of integrated oscillator techniques, it should

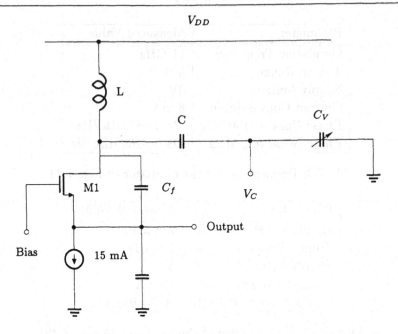

Figure 7.25: Colpitts oscillator with micromechanical capacitors

be noted that the design challenge posed by this category of circuits does not lie in their active part but in their passive elements. The architectures and circuit solutions are more or less known from discrete circuit design. The research effort is oriented towards designing and fabricating a better capacitor, varactor or inductor. The main criterion here is the enhancement of the quality factor, so that integrated VCOs can match the requirements set by modern telecommunications applications. In this process, the most critical element is the integrated inductor, since its quality factor usually falls short of that of the capacitor. By the exploitation of appropriate layout techniques and fabrication technology tweaking, it now seems feasible to realize integrated inductors with a Q higher than 10.

Bibliography

[1] B. Razavi, "Challenges in the Design of Frequency Synthesizers for Wireless Aplications," in *IEEE 1997 CICC , paper 20.1*, pp. 395–402.

[2] "Monolithic Phase - Locked Loops and Clock Recovery Circuits," in *IEEE Press , edited by B. Razavi*, 1996.

[3] M. Steyaert, *Wireless CMOS Receiver - Low Phase Noise Gigahertz VCOs in CMOS*, EPFL Lausanne Switcherland, 1995.

[4] B. De Smedt and G. Gielen, "Accurate Simulation of Phase Noise in Oscillators," in *ESSCIRC 97*, pp. 208–211.

[5] A.A.Abidi and R. G. Meyer, "Noise in Relaxation Oscillators," *IEEE Journal of Solid - State Circuits*, Dec. 1983.

[6] J.A. McNeill, *Jitter in Ring Oscillators*, Ph.D. thesis, Boston University, 1994.

[7] D.B. Leeson, "A Simple Model of Feedback Oscillator Noise Spectrum," in *Proc. of IEEE*, Feb. 1996.

[8] T.C. Weigandt, B. Kim, and P.R. Gray, "Analysis of Tuning Jitter in CMOS Ring Oscillators," in *Proc. IEEE ISCAS*, June 1994.

[9] B. Razavi, "Analysis, Modeling and Simulation of Phase Noise in Monolithic Voltage - Controlled Oscillators," in *Proc. IEEE CICC*, May 1995.

[10] A. Demir and A. Sangiovanni Vincentelli, "Simylation and Modelling of Phase Noise in Open - Loop Oscillators," in *Proc. IEEE CICC*, May 1996.

[11] M. Soyuer, "High - Frequensy Phase - Locked Loops in Monolithic Bipolar Technology," in *IEEE J. of Solid - state Circuits*, June 1989, vol. SC-24, pp. 787–795.

[12] T-P Liu and R.G. Meyer, "A 250 MHz Monolithic Voltage - Controlled Oscillator with Low Temperature Coefficient," in *IEEE J. of Solid - State Circuits*, April 1990, vol. 25, No 2, pp. 555–561.

[13] N. M. Nguyen and R.G. Meyer, "A 1.8 GHz Monolithic LC Voltage - Controlled Oscillator," in *IEEE J. of Solid - State Circuits*, March 1992, vol. 27, No 3, pp. 444–450.

[14] M. Soyuer, K.A. Jenkins, J.N. Burghartz, H.A. Ainspan, F.J. Canora, S. Ponnapalli, J. F. Ewen, and W.E. Pence, "A 2.4 GHz Silicon Bipolar Oscillator with Integrated Resonator," in *IEEE J. of Solid - State Circuits*, Feb. 1996, vol. 31, No 2, pp. 268–270.

[15] J. Granickx and M.S.J. Steyaert, "A 1.8 - GHz CMOS Low Phase - Noise Voltage - Controlled Oscillator with Prescaler," in *IEEE J. of Solid - State Circuits*, Dec. 1995, vol. 30, No 12, pp. 1474–1482.

[16] M. Soyuer, K.A. Jenkins, J.N. Burghartz, and M.D. Hulvery, "A 3-V 4-GHz nMOS Voltage - Controlled Oscillator with Integrated Resonator," in *IEEE J. of Solid - State Circuits*, Dec. 1996, vol. 31, No12, pp. 2042–2045.

[17] J. Granickx and M.S.J. Steyaert, "A 1.8 - GHz Low Phase - Noise CMOS VCO Using Optimized Hollow Spiral Inductors," in *IEEE J. of Solid - State Circuits*, May 1997, vol. 32, No5, pp. 736–744.

[18] A. Rofougaran, J. Rael, M. Rofougaran, and A. Adidi, "A 900 MHz CMOS LC-Oscillator with Quadrature Outputs," in *IEEE ISSCC96, paper 24.6*, pp. 392–393.

[19] B. Raravi, "A. 1.8 GHz CMOS Voltage - Controlled Oscillator," in *IEEE ISSCC97, paper 23.6*, pp. 388–389.

[20] M. Thamsirianunt and T.A. Kwasniewski, "CMOS VCO's for PLL Frequency Synthesis in GHz Digital Mobile Radio Communications," in *IEEE J. of Solid - State Circuits*, Oct. 1997, vol. 32, No10, pp. 1511–1524.

[21] T. Aytur and B. Razavi, "A 2GHz, 6mW BiCMOS Frequency Synthesizer," in *IEEE ISSCC95, paper 15.4*, pp. 264–265.

[22] M. Soyuer, J.N. Burghartz, et al., "An 11-GHz 3-V SiGe Voltage Controlled Oscillator with Intergrated Resonator," in *IEEE J. of Solid - State Circuits*, Sept. 1997, vol. 32, No 9, pp. 1451–1454.

[23] D.J. Young and B.E. Boser, "A Micromachine - Based RF Low - Noise Voltage - Controlled Oscillator," in *IEEE CICC 1997*, pp. 431–434.

[29] D.J. Young and B.E. Boser, "A Micromachine - Based RF Low - Noise - Voltage - Controlled Oscillator," in IEEE CICC 1997, pp. 431-434.

CHAPTER
8

INTEGRATED RF FILTERS

8.1 INTRODUCTION

A communications transceiver hosts an assortment of filters that cover a wide frequency span (from audio to RF) and vary greatly in performance, depending on their function. Some of these filters can be easily integrated (at low frequencies) while others have yet to be put on a silicon chip. This causes many problems in transceiver design since it increases the component count - and therefore costs - while it raises the requirements from the other circuits in the system, as will be shown later on. The main reasons that keep certain filters out of the chip include the high frequency of operation, the high required quality factor and the stringent specifications in terms of dynamic range and power dissipation. In Fig. 8.1, the receiver part of a super-heterodyne system is shown.

The preselect filter is placed directly after the antenna and before the

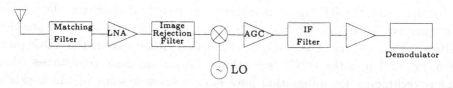

Figure 8.1: Super-heterodyne receiver schematic

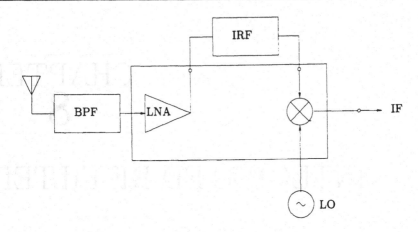

Figure 8.2: Integrated RF circuit

LNA input. Operating in radio frequencies with an extremely high Q and at a prohibitive SNR, this filter cannot be integrated for practical use. Following the LNA output, an image-reject filter precedes the mixer to prevent the image frequency from corrupting the IF mixer output that carries the desired signal. Image-reject filters remain also off-chip in industrial implementations (although there has been some effort to integrate them, as will be shown later in this chapter). This leads to the inconvenience depicted in Fig. 8.2: even though the LNA and mixer cells may reside on the same chip, the LNA output signal has to be driven off-chip, through an external image-reject filter (IRF) and then into the chip again, before it is fed to the mixer RF input. This means that the LNA output and the mixer input should be matched to 50Ω, which poses an extra burden on their respective specifications (e.g. linearity - power dissipation).

The IF filter, depending on the IF band used in the application, is generally more easily integrated. The methods adopted to address the problem described above are two: a) using an architecture other than super-heterodyne to eliminate the need for such filters, and b) designing and implementing integrated RF filters.

Fig. 8.3 shows the simplified schematic of a modern digital transceiver with a direct conversion architecture. In direct conversion there is no IF frequency and the RF signal is directly converted to baseband data. This circumvents the use of high frequency filters, except for the baseband preselect filters at the antenna. The baseband filters that either precede the ADCs (anti-aliasing) or follow the DACs (smoothing) operate in video frequencies where many techniques for integration have been proposed, some of which exhibit

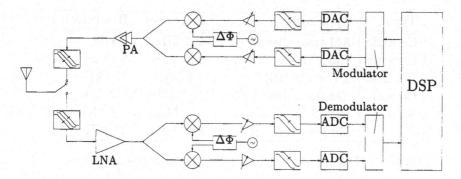

Figure 8.3: Direct conversion architecture

impressive performance in terms of power dissipation (e.g. [1-2]).

In this chapter, an attempt will be made to record the efforts to integrate RF filters. This is an on-going research, still in its first steps (1998).

The alternatives in filter integration are the following:

- Digital filters.

- Passive analog filters.

- Active analog filters (Gm -C, MOSFET - C).

- Switched-capacitor filters.

- LC filters with active Q-enhancement circuitry.

Depending on the exact application, some of the above options are excluded due to overwhelming technical hurdles. The latter case (LC filters with active Q-enhancement circuitry) is a novel, promising technique. It will be thoroughly investigated soon after.

On the other hand, the hitherto utilized discrete filters involve ceramic, crystal, Surface Acoustic Wave (SAW) and LC filters. These filters generally exhibit small bandwidth (high selectivity), at zero power consumption and low cost. This low cost is the main reason that these filters have not yet been replaced by integrated ones: The silicon real estate that the integrated counterparts would occupy should be adequately small to make their cost competitive to discrete.

Table 8.1 [3] summarizes the basic characteristics from a series of integrated filters for telecommunications applications, so that the reader can have an overview of the present state-of-the-art.

Type	Application	Frequency	Bandwidth
Ceramic	AM transmission (IF)	262 kHz	6 kHz
Ceramic	Pager (IF)	450 kHz	6 kHz
Ceramic	TV audio (IF)	4.5 MHz	120 kHz
Ceramic	FM transmission (IF)	10.7 MHz	230 kHz
SAW	DECT (IF)	110 MHz	1.1 MHz
LC	Mobile phone (RF)	881 MHz	25 MHz

Table 8.1: Discrete filters for telecommunications applications

8.2 INTEGRATED FILTERS IMPLEMENTATION OVERVIEW

a) **Digital Filters** One would reasonably expect that a fast, modern DSP would implement any filter up to several hundred MHz. However, in practice the maximum attainable frequency of operation does not exceed 1 MHz. This is partly accounted to the fact that fast and highly accurate ADCs, as well as analog anti-aliasing filters are required, leading to considerably large silicon area and power dissipation. The latter is an especially annoying problem, since the consumed power rises with frequency.

Finally, one should not overlook the problem of crosstalk between the digital part and the analog circuits, when these share the same silicon chip. This calls for careful layout, following certain guidelines for crosstalk minimization.

b) **Passive *LC* Filters** It is a well-known fact that the inductor is the single most challenging element to integrate on silicon. As it was mentioned in Chapter 3, considerable effort has been dedicated lately to integrated inductor modeling and many reported RF circuits utilize on-chip inductors. However, the typically low value of quality factor has prevented the usage of on-chip inductors in integrated *LC* networks. This problem is not so pronounced in GaAs processes where inductors have been extensively used in MMICs for years. The utilization of alternative metallization materials in silicon processes (e.g. gold over passivation) will enable the fabrication of high Q inductors, at a cost premium. This could pave the way for passive integrated filter design.

c) **Passive Electroacoustic Filters** Electroacoustic filters can be integrated, provided that certain process modifications are made [4-5]. The reported results are very encouraging, since the produced filters exhibit zero power consumption and very good characteristics. On the downside, the required silicon real estate is excessive. The main categories in this case are BAW (bulk acoustic wave) and SAW (surface acoustic wave) filters.

BAW filters are laid out on four layers: a layer of gold is deposited on a SiO_2 substrate, followed by a piezoelectric ZnO layer. The uppermost layer is aluminum. This process is supposedly compatible with the fabrication of active elements on silicon, thus leading to acceptable fabrication costs. SAW filters are made in an analogous way. A detailed presentation of this category of filters would escape the scope of this book.

d) Switched-Capacitor Filters   Switched-capacitor filters provide a good alternative at lower frequencies (up to 10 MHz). Among their disadvantages are their relatively high power consumption and the fact that they are essentially discrete time systems.

e) Integrated Continuous-Time Analog Filters   These active filters have received special focus in the last years. Many different design techniques have been reported, that address several applications. The main categories are Gm - C, MOSFET - C, OTA - C and OTA - MOSFET - C filters. In [6], an extensive review of these techniques can be found, demonstrating their respective advantages and disadvantages.

The most recent applications of the above techniques involve filters that operate in the vicinity of 100 MHz (disk-drive read channel applications). The frequency of operation is expected to rise in the near future. However, for continuous-time analog filters in the RF band the use of integrated inductors is dictated.

A broad application domain for the above techniques is video-frequency filtering that includes the ubiquitous analog filters that precede ADCs or follow DACs.

The main shortcomings of active continuous-time analog filters are their relatively high power dissipation (under conditions), the considerable silicon area they usually occupy, and their moderate dynamic range. It should be noted that the bulk of different implementations and applications do not allow for a concise categorization in terms of performance. For a full and systematic presentation the reader is referred to [7].

f) LC Filters with Active Q-Enhancement   The main drawback with integrated inductors is their low quality factor. This is mainly attributed to the ohmic losses of the metal that is used for laying out the spiral turns. These losses can be simply modeled by connecting to the inductance a series resistor

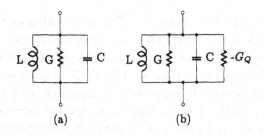

Figure 8.4: Resonant circuit model

(or its shunt equivalent). Therefore, if a negative resistance is connected to the LC tank in such a way that the ohmic losses are compensated, the integrated LC circuit approaches an ideal (i.e. lossless) resonant element. This negative resistance can be implemented with active elements. Recently, some techniques have been proposed that lead to integrated LC filters with active Q-enhancement that operate in the area of a few GHz. In the rest of this chapter these techniques will be presented in detail.

8.3 LC FILTERS WITH ACTIVE Q-ENHANCEMENT

8.3.1 Integrated Second Order RLC Filter at 1.8 GHz

In [8], a second order filter (biquad) is presented, providing quality factor and center frequency tuning. The filter's response is bandpass, centered around 1.8 GHz. It is implemented in a bipolar process.

To illustrate the principle of operation for the active negative resistance circuit in the filter, Fig. 8.4(a) presents a parallel RLC resonant element (LC tank where R corresponds to the inductor's ohmic losses). In the figure, the resistance R is represented by a conductance G connected in parallel to the LC tank. In Fig. 8.4(b), a negative conductance $-G_Q$ is connected in parallel to the resonant circuit.

If this negative conductance becomes equal in magnitude to the conductance of inductor losses, the two are cancelled out and an ideal LC resonant element is yielded. Essentially, the negative resistance active circuit replaces the energy dissipated by the ohmic losses on the inductor. Under the condition that $G - G_Q > 0$, the Q of the circuit is finite and positive and can be tuned by adjusting the value of G_Q.

The center frequency of the filter can be tuned independently of the quality factor. A well-known method employs an array of capacitors that can be

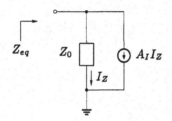

Figure 8.5: Impedance multiplication technique

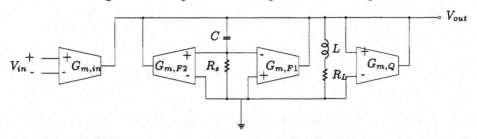

Figure 8.6: Second order bandpass RLC filter

selectively connected to the circuit through switches, thus realizing a variable capacitance. Alternatively, a varactor can be used, if the technology allows for such an element to be fabricated with low losses and acceptable linearity.

Another technique that can be employed for center frequency tuning is known as impedance multiplication and is depicted in Fig. 8.5. The passive element utilized for the tuning has an impedance Z_0. A current source, controlled by the current I_2 that flows through Z_0, is connected in parallel to the passive element. The controlled source exhibits a gain A_I. The resulting equivalent input impedance is given by the following equation:

$$Z_{eq} = \frac{Z_0}{A_I + 1} \qquad (8.1)$$

Consequently, by electronically adjusting the gain A_I of the controlled source the equivalent impedance Z_{eq} of the combination is adjusted, thus providing center frequency tuning for the filter.

The schematic of Fig. 8.6 displays a second order bandpass filter circuit, with center frequency and quality factor tuning. The inductor used, L, is shown connected in series to the parasitic ohmic loss resistor R_L. The transconductor $G_{m,Q}$ is a two terminal device exhibiting a conductance equal to $-G_{m,Q}$. Therefore, the required negative conductance is attained and the filter Q can be tuned by adjusting the value of $G_{m,Q}$.

The element Z_0 in the circuit is implemented by the capacitor C. The

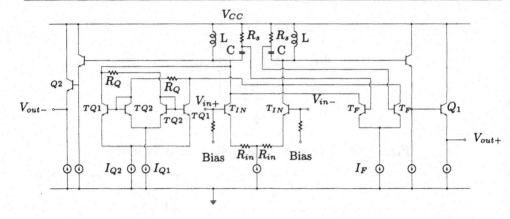

Figure 8.7: Complete balanced form of Fig. 8.6

resistor R_s senses its current and the resulting voltage drop is converted to current by the transconductors $G_{m,F1}$ and $G_{m,F2}$. The equivalent current gain A_I is $(G_{m,F1} - G_{m,F2})R_s$. Hence, by controlling $G_{m,F1}$ and $G_{m,F2}$ the current gain can be adjusted to positive as well as negative values. Finally, $G_{m,in}$ is the input transconductor that feeds current to the resonant LC circuit.

The topology in Fig. 8.6 can be implemented by the circuit in Fig. 8.7 in balanced form.

Transistors T_{Q_1}, T_{Q_2} and resistors R_Q implement the transconductor $G_{m,Q}$ of Fig. 8.6. The resistors act as voltage-to-current converters and the transistors amplify the current by a gain controlled by bias currents I_{Q_1} and I_{Q_2}. This technique achieves a broad range of transconductance tuning and therefore filter Q tuning. $G_{m,Q}$ is given by the following formula:

$$G_{m,Q} = \frac{g_{mQ_1} - g_{mQ_2}}{1 + g_{mQ_2}R_Q} \tag{8.2}$$

where g_{mQ_i} are the transconductances of the respective transistors.

Transistors T_F implement the transconductor $G_{m,F1}$ of Fig. 8.6. The other transconductor, $G_{m,F2}$, is not shown in Fig. 8.7. Transconductor $G_{m,F1}$ is controlled by the bias current I_F. The input transconductor $G_{m,in}$ is implemented by the differential pair of transistors T_{IN}, in emitter degenerated configuration through resistors R_{IN}. Finally, to drive the load at the outputs $(V_{out}+, V_{out}-)$, two pairs of common emitter transistors are connected in series.

The circuit has been fabricated in a bipolar 0.8 μm process, where the npn device has an f_τ value of 25 GHz. All the inductors and capacitors are on chip. The active part of the filter occupies $0.38 mm^2$ of silicon area. The main

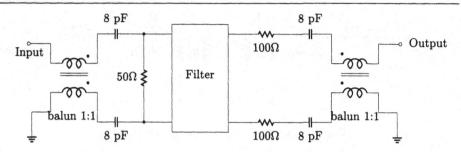

Figure 8.8: Experimental setup for the filter in Fig.8.7

Parameter	Measured Value
Supply Voltage	2.8 ÷ 4.5 V
Current Dissipation	
(without output buffers)	8.7 mA
Q Tuning Range	3 ÷ 350
Frequency Tuning Range	1.6 ÷ 2 GHz
Gain	7 dB
Dynamic Range for 1 dB CP	40 dB
Spurious-Free Dynamic Range (SFDR)	30 dB
OIP3	-9 dBm

Table 8.2: Performance of the integrated filter in Fig. 8.7

experimental results are summarized in Table 8.2 and were obtained using the measurement setup shown in Fig. 8.8.

8.3.2 Integrated Bandpass LC Filter at 200 MHz

A similar LC Q-enhancement technique appears in [9], where a fourth order bandpass filter is designed, centered at 200 MHz with $Q = 100$. This filter is intended for IF filtering in modern mobile telephony receivers.

In this case, a coupled resonators topology was followed [10], as it is shown in Fig. 8.9.

The schematic displays two buffers g_{mi} and g_{mo}, at the input and at the output respectively, and two magnetically coupled resonant RLC elements. The resistors R_1 and R_2 correspond to the combination of ohmic losses and Q-enhancement negative resistance elements for inductors L_1 and L_2.

The output voltage can be expressed as a function of currents I_1 and I_2 and inductances L_1 and L_2:

$$V_2 = sL_2I_2 + sMI_1 \tag{8.3}$$

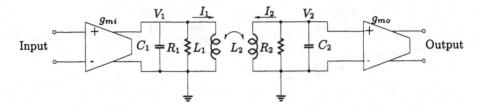

Figure 8.9: Coupled resonators topology

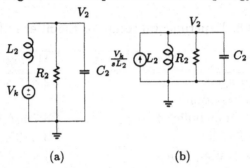

(a)   (b)

Figure 8.10: Equivalent magnetic coupled resonant circuits

where $M = k\sqrt{L_1 L_2}$. Assuming that the two resonant circuits are identical, therefore $L_1 = L_2 = L$, it is found that $M = kL$. The Thevenin and Norton equivalent circuits for the second resonant circuit ($L_2 C_2 R_2$) are shown in Figs. 8.10(a) and 8.10(b) respectively. The voltage V_k can be derived from Eq. (8.3):

$$V_k = skLI_1 \tag{8.4}$$

for the case where $L_1 = L_2 = L$. For high Q resonant circuits, $M \ll L$ yielding $I_1 \approx \frac{V_1}{sL}$ thus

$$V_k \approx kV_1 Z(s) \tag{8.5}$$

From Fig. 8.10(b) it is derived that:

$$V_2 \approx \frac{kV_1}{sL_2} Z(s) \tag{8.6}$$

where $Z(s)$ is the impedance of the parallel RLC resonant circuit. Similarly, for V_1:

$$V_1 \approx g_{mi} V_i Z(s) + \frac{kV_2}{sL_1} Z(s) \tag{8.7}$$

Equations (8.6) and (8.7) can be solved to provide the transfer functions of interest, V_1 vs. V_i and V_2 vs. V_i. A bandpass response is thus attained. The

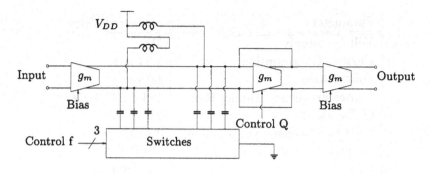

Figure 8.11: Second order stage schematic

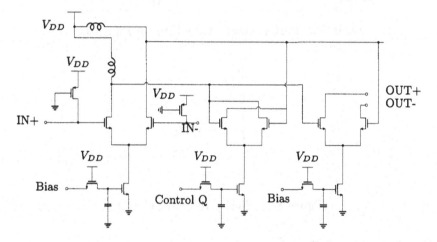

Figure 8.12: Second order stage circuit schematic

schematic of the active filter that was designed according to this technique is shown in Fig. 8.11.

This is one of the two second order stages in fully balanced form. The appropriate inductor for fully balanced operation is laid out on silicon as a 26-turn center-tapped spiral. This inductor has a nominal value of 250 nH and a self-resonant frequency of 270 MHz. By employing an array of capacitors where each one is controlled by a switch, the center frequency of the filter can be adjusted by means of a digital logic. The detailed schematic of the second order stage is demonstrated in Fig. 8.12, which includes the input and output buffers, the load comprising the center-tapped spiral, and the active implementation of the negative resistance.

The above circuit was implemented in a CMOS 2 μm process. The results are summarized in Table 8.3.

Parameter	Measured Value
Voltage Supply	3V
Current Consumption	2.94 mA
Silicon Area	1.7mm^2
Center Frequency Tuning Range	194 ÷ 203 MHz
Q Tuning Range	2.3 ÷ 100
3 dB Bandwidth (typical value)	2 MHz
Gain (with matching)	-5 dB
CP1	-12 dBm
DR	47 dB
SFDR	37 dB

Table 8.3: Performance of the filter of Fig. 8.11

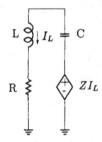

Figure 8.13: Series resonant circuit

8.3.3   Integrated Bandpass RLC Filter at 1 GHz

Based on the above, the loss resistor of the actual inductor is connected in parallel to a negative resistance for Q-enhancement. In this case, the resonant circuit is a parallel RLC. Its dual form is shown in Fig. 8.13. This is a series resonant RLC topology, where a current-controlled voltage source is connected in series as well. This controlled source is implemented by a transresistance amplifier that detects the current flowing through the inductor and outputs a suitable voltage for Q-enhancement. If the transresistance gain Z is real, the resonance frequency is not affected and the equivalent Q becomes:

$$Q = \frac{Q_{\text{inductor}}}{1 - \frac{Z}{R}} \tag{8.8}$$

An active implementation of the technique of Fig. 8.13 is presented in [11]. The respective schematic is shown in Fig. 8.14.

Despite the fact that the input signal can be inserted in any node of the

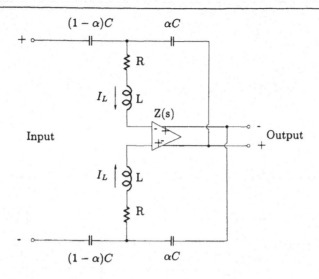

Figure 8.14: Implementation of the topology of Fig. 8.13

circuit, in this implementation it is fed to the capacitor, so that a bandpass response is obtained at the low resistance output of the transresistance amplifier. The transfer function of the circuit in Fig. 8.14 is given by Eq. (8.9).

$$G(s) = \frac{(1-\alpha)CR_{dc}s}{e(s)} \tag{8.9}$$

where R_{dc} is the dc gain of the transresistance amplifier which exhibits a single-pole response:

$$Z(s) = \frac{CR_{dc}}{1 + s\tau} \tag{8.10}$$

The polynomial $e(s)$ is given by:

$$e(s) = 1 + s[C(R - aR_{dc}) + \tau] + s^2(LC + \tau CR) + s^3 LC\tau \tag{8.11}$$

The circuit implementing the schematic of Fig. 8.14 is shown in Fig. 8.15.

The circuit comprises a Gilbert cell ($Q_1 - Q_4$) driving an ohmic load. The two filter outputs pass through buffers formed by emitter followers Q_5 and Q_6. The circuit employs two 5 nH inductors with a Q of 2 at 1 GHz. The capacitors have nominal values of 1 pF, with $\alpha = 0.6$. Quality factor tuning is accomplished through V_Q or V_{B2}.

In this particular implementation there is no provision for center frequency tuning. The circuit has been fabricated in 0.8 μm BiCMOS. Measurements

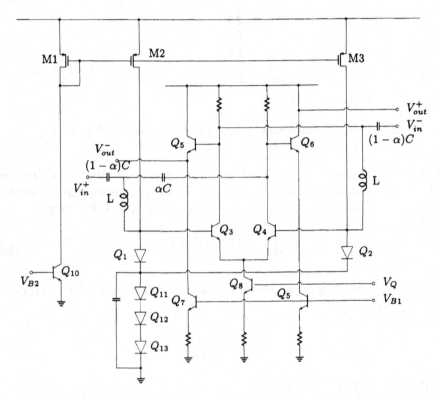

Figure 8.15: The bandpass filter circuit of Fig. 8.14

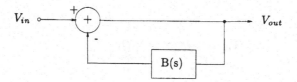

Figure 8.16: Closed-loop system

have revealed a Q of 494 at 740 MHz. A drawback of this topology is its low linearity: for an input power of -40 dBm, the signal-over-distortion ratio was a mere 36 dB.

8.4   NOTCH FILTER FOR IMAGE REJECTION

Recently [12], an integrated RF receiver for applications at 1.9 GHz was presented, including a tunable integrated notch filter for image rejection. This is the first reported attempt at an integrated image rejecting system that does not make use of an image-reject mixer. The attained rejection in the mixer case is between 30 and 45 dB. The notch filter technique presented below achieves an image rejection better that 50 dB. The principle of operation of the notch filter is given below. The transfer function of the system in Fig. 8.16 is:

$$H(s) \equiv \frac{V_{out}}{V_{in}} = \frac{1}{1 + B(s)} \tag{8.12}$$

if $B(s)$ is a second order bandpass function,

$$B(s) = \frac{G_{BP}\frac{\omega_o}{Q}s}{s^2 + s\frac{\omega_o}{Q} + \omega_o^2} \tag{8.13}$$

then $H(s)$ becomes

$$H(s) = \frac{s^2 + s\frac{\omega_o}{Q} + \omega_o^2}{s^2 + s\frac{\omega_o}{Q}(1 + G_{BP}) + \omega_o^2} \tag{8.14}$$

Expression (8.14) yields a pair of zeros at a frequency ω_o, with a quality factor Q. These zeros produce a notch at ω_o and the system of Fig. 8.16 becomes a notch filter. For $\omega = \omega_o$, this leads to

$$H(s)\,|_{\omega=\omega_o} = \frac{1}{1 + G_{BP}} \tag{8.15}$$

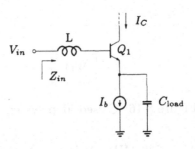

Figure 8.17: Series resonant LC circuit

If $G_{BP} \gg 1$ the filter is very efficient in terms of input image rejection. It is thus required to implement $B(s)$ with a sufficiently high dc gain. The circuit of choice in this case is a series resonant LC shown in Fig. 8.17. This is a variation of the Colpitts common collector oscillator: An integrated inductor is connected in series to the base of a transistor in emitter follower configuration, where the emitter drives a capacitive load. The resonant frequency ω_o for the LC is given by:

$$\omega_o = \frac{1}{\sqrt{LC_s}} \tag{8.16}$$

where C_s is the series combination of C_π - the capacitance of the bipolar transistor Q_1 - and C_{load}. If the ohmic losses of the inductor are represented by R_s, the input impedance Z_{in} of the circuit is given by Eq. (8.17):

$$Z_{in} = j\omega L = \frac{1}{j\omega}\left(\frac{1}{C_\pi} + \frac{1}{C_{load}}\right) + r_s + r_b - \frac{g_m}{\omega^2 C_\pi \cdot C_{load}} \tag{8.17}$$

where r_b is the base resistance of the bipolar transistor and g_m is its transconductance. The negative term in Eq. (8.17) represents the negative resistance encountered when looking at the base of Q_1. Its value depends on the transconductance of the transistor and thus can be adjusted by the bias current I_b of the emitter follower. Therefore, Q tuning of the series resonant LC can be effected through I_b. This circuit is employed in a feedback loop, as shown in Fig. 8.18, to obtain a notch filter: The input voltage V_{in} is converted to a current (I_{C2}) which is in fact the input current. The emitter current of Q_3 (I_{E3}) is the output current. Node A is the summation point of the schematic of Fig. 8.16. Current I_s is the input current of the resonant circuit.

The circuit has been fabricated in a 0.5 μm bipolar process, where the npn element has an f_τ of 25 GHz. The inductor is laid out on the upper metallization layer and the capacitors are of the MIM type. The measurements

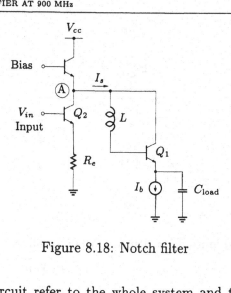

Figure 8.18: Notch filter

supplied for this circuit refer to the whole system and therefore individual characteristics of the filter cannot be extracted. Still, there are some image rejection measurements for various filter Q values. At an IF frequency of 300 MHz, image rejection varies between 41.5 dB and 71.5 dB - depending on the Q.

8.5   CMOS BANDPASS AMPLIFIER AT 900 MHz

A band select filter after the antenna and before the LNA input is a prerequisite in every modern mobile communications transceiver system. For example, the channel bandwidth at 900 MHz is 25 to 35 MHz, resulting in a required filter Q of 30. On top of that, a low noise figure and a high linearity are necessary, making the integration of the band select filter a very difficult task. A viable solution in this case could be the combination of the filter and the amplifier in one circuit, an RF tuned amplifier. The basic circuit topology of a tuned amplifier is given in Fig. 8.19. This is a cascode stage with a parallel LC load, followed by an output buffer that drives a load of 50 Ω. The gain of the circuit is given by:

$$A_v = \frac{-g_{m1}}{G_{\text{load}}} A_{\text{buffer}} \qquad (8.18)$$

where g_{m1} is the transconductance of transistor M1, G_{load} is the conductance of the resonant circuit and A_{buffer} is the voltage gain of the buffer. It is obviated from (8.18) that the gain depends on the quality of the LC resonance. The main issue therefore is the quality of integrated inductors on Si processes.

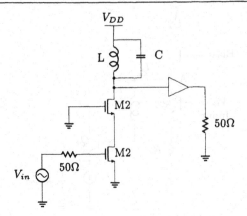

Figure 8.19: Tuned CMOS amplifier

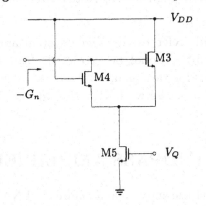

Figure 8.20: Negative conductance circuit

In [13], inductor quality factor enhancement is proposed for tuned amplifiers, through the introduction of the negative conductance circuit of Fig. 8.20. The latter is a positive feedback circuit, exhibiting a negative conductance given by (8.19):

$$-G_n = -\frac{g_{m3}g_{m4}}{g_{m3} + g_{m4}} \qquad (8.19)$$

The conductances of transistors M_3 and M_4 are adjusted through the bias current of transistor M_5 and therefore the Q of the amplifier is tuned through V_Q. In order to provide frequency tuning to the amplifier, the use of a Miller capacitor is proposed, to modify the overall capacitance of the LC. The employed circuit technique is shown in Fig. 8.21.

This is a variable gain amplifier stage with a capacitor C_F connected in

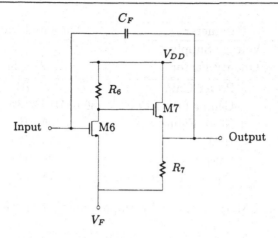

Figure 8.21: Center frequency tuning for the amplifier

feedback. The amplification gain $-A_f$ is

$$-A_f = g_{m6}R_6K_7 \tag{8.20}$$

where K_7 is the gain of the source follower M_7. The value of the transconductance of transistor M_6 is adjusted through voltage V_F and gain control is thus achieved. The Miller capacitor appearing at the input of the circuit of Fig. 8.21 due to C_F, is:

$$C_{eff} = (1 + A_f)C_F \tag{8.21}$$

Concluding, the overall circuit comprises a tuned amplifier (Fig. 8.19), a Q-enhancement circuit (Fig. 8.20), a center frequency tuning circuit (Fig. 8.21) and an output buffer. The circuit has been designed and fabricated in a 0.8 μm CMOS process. The experimental results are summarized in Table 8.4.

8.6 PASSIVE INTEGRATED FILTERS

The progress in silicon technologies in the latest years, gradually allows for conventional RLC filters to be laid out on a silicon substrate. If this is ultimately achieved in an easy and systematic fashion, it will pave the way for the complete integration of transceivers with high reliability, better performance and lower consumption. The two main problems towards the integration of passive filters involve the silicon area required for passive elements (capacitors and inductors) and the lack of tunability that would compensate for process skews that affect the filter's transfer function. In [14], one of the first attempts

Parameter		Measured Value
Voltage Supply		3V
Quality Factor		2.2 ÷ 44
	Power Gain	17 dB
	Center Frequency	869 MHz ÷ 893 MHz
Q = 30	Noise Figure	6 dB
	IIP3	-14 dBm
	Output CP1	-30 dBm
	Power Consumption	78 mW

Table 8.4: CMOS Integrated bandpass amplifier performance

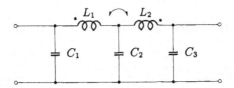

Figure 8.22: Passive fifth order lowpass LC filter

at silicon integrated passive filters is presented. This is a fifth order maximally flat lowpass LC filter, depicted in Fig. 8.22. The filter has been fabricated in a typical bipolar process where the *npn* element has an f_τ of 8 GHz. The 3 dB frequency of the filter is 880 MHz.

It should be noted that a careful modeling of the passive elements yielded simulation results very close to the measured testchip results. Some indicative values are: $L_1 = L_2 = 9.7nH$, $C_1 = C_3 = 1.3pF$, $C_2 = 4.3pF$. Measurement and simulation have revealed that the inductors are only slightly affected by process and geometry tolerances and that a good matching between the capacitors minimizes the effect of their absolute value on the filter's response.

The use of better and more accurate models for integrated passive elements, such as the ones presented in Chapter 3, will permit the reliable and efficient design of integrated filters in the future.

Bibliography

[1] R.H. Zele and D.J. Allstot, "Low - Power CMOS Continuous - Time Filters," *IEEE J. Solid-State Circuits*, vol. 31, pp. 157–168, Feb. 1996.

[2] S. Bantas and Y. Papananos, "A W - power continuous - time current - mode filter in a digital CMOS proccess," *IEEE 5th Int. Conf. on VLSI and CAD*, pp. 346–348, Seoul - Korea 1997.

[3] W. B. Kuhn, *Design of Integrated, Low Power, Radio Receivers in BiC-MOS Technologies*, Ph.D. thesis, Virginia Polytechnic Institute, Dec. 1995.

[4] J. C. Haartsen, "Development of a Monolithic Programmable SAW Filter in Silicon," *IEEE MTT-S Digest*, pp. 1115–1118, 1990.

[5] P. T. M. van Zeijl, J. H. Visser, and L. K. Nanver, "FM Radio Receiver Front - End Circuitry with On - Chip SAW Filters," *IEEE Trans. on Consumer Electr.*, vol. 35, no. 3, Aug. 1989.

[6] Y. P. Tsividis, "Integrated Continuous - Time Filter Design - An Overview," *IEEE J. Solid-State Circuits*, vol. 29, no. 3, pp. 166–176, Mar. 1994.

[7] "Integrated Continuous - Time Filters," *IEEE Press, Edited by Y. P. Tsividis and J. O. Voorman*, 1992.

[8] S. Pipilos, Y. P. Tsividis, J. Fenk, and Y. Papananos, "A Si 1.8 GHz RLC Filter with Tunable Center - Frequency and Quality Factor," *IEEE J. Solid-State Circuits*, vol. 31, no. 10, pp. 1517–1525, Oct. 1996.

[9] W. B. Kuhn, W. Stephenson, and A. Elshabini Riad, "A 200 MHz CMOS Q-Enhanced LC Bandpass Filter," *IEEE J. Solid-State Circuits*, vol. 31, no. 8, pp. 1112–1122, Aug. 1996.

[10] H. L. Krauss, C. W. Bastian, and F. H. Raab, *Solid State Radio Engineering*, Wiley, New York, 1980.

[11] R. Dancan, K. W. Martin, and A. S. Sedra, "Q-Enhanced Active - RLC Bandpass Filter," *IEEE Trans. on Circuits and Systems - II*, vol. 44, no. 5, pp. 341–347, May 1997.

[12] J. A. Macedo and M. A. Copeland, "A 1.9 - GHz Silicon Receiver with Monolithic Image Filtering," *IEEE J. Solid-State Circuits*, vol. 33, no. 3, pp. 378–386, March 1998.

[13] C. Y. Wu and S. Y. Hsiao, "The Design of a 3-V 900 MHz CMOS Bandpass Amplifier," *IEEE J. Solid-State Circuits*, vol. 32, no. 2, pp. 159–168, Feb. 1997.

[14] N. M. Nguygen and R. G. Meyer, "Si IC - Compatible Inductors and LC Passive Filters," *IEEE J. Solid-State Circuits*, vol. 25, no. 4, pp. 1028–1031, Aug. 1990.

List of Figures

List of Tables